Lectures in Mathematics
ETH Zürich
Department of Mathematics
Research Institute of Mathematics

Managing Editor:
Helmut Hofer

Charles M. Newman
Topics in
Disordered Systems

Birkhäuser Verlag
Basel · Boston · Berlin

Author's address:

Courant Institute of Mathematical Sciences
251 Mercer Street
New York, NY 10012
USA

1991 Mathematical Subject Classification 82B44, 82D30

A CIP catalogue record for this book is available from the
Library of Congress, Washington D.C., USA

Deutsche Bibliothek Cataloging-in-Publication Data
Newman, Charles M.:
Topics in disordered systems / Charles M. Newman. - Basel ; Boston ;
Berlin : Birkhäuser, 1997
    (Lectures in mathematics : ETH Zürich)
    ISBN 3-7643-5777-0 (Basel ...)
    ISBN 0-8176-5777-0 (Boston)

© 1997 Birkhäuser Verlag, P.O. Box 133, CH-4010 Basel, Switzerland
Printed on acid-free paper produced from chlorine-free pulp. TCF ∞
Printed in Germany
ISBN 3-7643-5777-0
ISBN 0-8176-5777-0

9 8 7 6 5 4 3 2 1

Contents

Preface

This book consists of lecture notes for a nachdiplom course on disordered systems given at the mathematics department of ETH-Zürich in 1996. There were eleven lectures, each consisting of two 45 minute parts. The first lecture took place on April 18 and the last on July 5. Mario Wütrich assisted in the preparation of these notes with excellent editing and word processing, for which I express my gratitude.

In the physics literature, the term disordered system refers to a statistical mechanics model in a random environment. My aim is not to survey the entire field but rather to focus on a narrowly chosen class of disordered systems — Ising models whose nonzero (pair) interactions are i.i.d. random variables. For these systems, I concentrate entirely on equilibrium statistical mechanics; there is no discussion of dynamics.

The two main types of systems considered here are disordered ferromagnets and spin glasses, with the latter both more interesting and more difficult. The emphasis is on questions concerning the number of equlibrium phases of such systems, i.e., the number of ground states (at zero temperature) or the number of extremal Gibbs states (at nonzero temperature).

A recurring theme in the first three chapters of these notes is that these questions about disordered Ising models are often closely connected to interesting issues concerning related percolation models. The percolation models that arise in this way are first-passage percolation (and related random minimal surfaces in higher dimensions), invasion percolation and standard Bernoulli bond percolation (along with its close relative, the dependent percolation of Fortuin-Kasteleyn random cluster models).

The fourth chapter of these notes concerns the low temperature behavior of the Edwards-Anderson model, a paradigm short-range spin glass. The question discussed here is whether and in what sense Parisi's analysis of the Sherrington-Kirkpatrick spin glass, a paradigm mean-field (or "infinite-range") model, is relevant to short-range spin glasses. Closely related is the more general conceptual issue of how to approach the thermodynamic (i.e., infinite volume) limit in systems which may have many complex competing phases. This issue has been addressed in recent joint work with Dan Stein and the main purpose of Chapter 4 is to give a mathematically coherent presentation of our approach.

These notes have benefitted from the comments of many people. In particular, Aernout van Enter read through much of the manuscript and suggested a number of improvements and Alain Sznitman raised the issue of measurability treated in

Appendix A. Some of the research reported here has been supported in part by NSA Grant MDA 904-96-1-0033 and by NSF Grant DMS-95-00868.

I thank Alain Sznitman for the invitation to visit the FIM at ETH, Helmut Hofer for the invitation to give a nachdiplom course and Ruth Ebel for her efficient help and general good cheer. Thanks are due all the faculty, postdocs, students and visitors who attended the course for their interest in the material, useful questions and patience as I learned how to erase the blackboards. For helping make my visit to Zürich so enjoyable, I thank the Ebel and Sznitman families for their hospitality. For keeping my bicycle over many years and transporting it from Marseille to Mons to Delft to Cambridge to Zürich, I thank Alain Messager, Jean Ruiz, Joël De Coninck, Mathew Harris, Alberto Gandolfi and Geoffrey Grimmett. Finally, I thank Arlene, Jennifer and Serena, my wife and daughters, for good-humoredly putting up with my lengthy absence from New York so that I could be in Zürich writing these notes.

New York City

March 4, 1997

Chapter 0

Introduction

Let Λ be a finite subset of $\mathbb{Z}^d$, for example, $\Lambda = \Lambda_L = \{-L, -L+1, \cdots, L\}^d$. An Ising model on Λ is a family $\{S_x : x \in \Lambda\}$ of random variables (called spins) taking values $+1$ or -1, whose joint distribution $P_{\Lambda,\beta}$ depends on a parameter $\beta \geq 0$ and has the form,

$$P_{\Lambda,\beta}(\{s\}) = Z_{\Lambda,\beta}^{-1} \exp\{-\beta H_\Lambda(s)\}. \tag{0.1}$$

Here $s \in \mathcal{S}_\Lambda = \{-1, +1\}^\Lambda$, $H_\Lambda(s)$ is a real valued function on $\mathcal{S}_\Lambda$, called the (finite volume) Hamiltonian, and $Z_{\Lambda,\beta}$, called the partition function, is a constant, such that $P_{\Lambda,\beta}(\mathcal{S}_\Lambda) = 1$. We call $P_{\Lambda,\beta}$ the (finite volume) Gibbs distribution or Gibbs measure or Gibbs state and β the inverse temperature.

We will primarily focus on Hamiltonians such as

$$H_\Lambda(s) = H_\Lambda^f(s) = \sum_{\substack{x,y \in \Lambda \\ \{x,y\} \in \mathbb{E}^d}} -J_{\{x,y\}}^\Lambda s_x s_y, \tag{0.2}$$

where $\mathbb{E}^d$ is the set of nearest neighbor bonds (or edges) $\{x, y\}$ of $\mathbb{Z}^d$, i.e., pairs of sites (or vertices) of $\mathbb{Z}^d$ whose Euclidean distance $||x - y||$ equals 1. The real valued $J_{\{x,y\}}^\Lambda$'s are called couplings. The superscript f is for "free", denoting the absence of boundary conditions (b.c.'s). The simplest b.c. (called a fixed b.c.) is the specification of a spin configuration, $\bar{s} \in \{-1, +1\}^{\partial\Lambda}$, on the boundary of Λ,

$$\partial\Lambda = \{y \in \mathbb{Z}^d \setminus \Lambda : \{x, y\} \in \mathbb{E}^d \text{ for some } x \in \Lambda\}, \tag{0.3}$$

and the replacement of (0.2) by

$$H_\Lambda(s) = H_\Lambda^{\bar{s}}(s) = H_\Lambda^f(s) - \sum_{\substack{x \in \Lambda, y \in \partial\Lambda \\ \{x,y\} \in \mathbb{E}^d}} J_{\{x,y\}}^\Lambda s_x \bar{s}_y. \tag{0.4}$$

The corresponding Gibbs measures will be denoted $P_{\Lambda,\beta}^f$ and $P_{\Lambda,\beta}^{\bar{s}}$.

A (finite volume) ground state is any spin configuration $s \in \mathcal{S}_\Lambda$, which minimizes the Hamiltonian $H_\Lambda(s)$. It should be clear that the limit as β tends to infinity (corresponding to zero temperature) of $P_{\Lambda,\beta}$ is the uniform measure on the ground

1

states of H_Λ. We will be interested in the infinite volume limits of ground states and Gibbs distributions, i.e., the limits as $\Lambda \to \mathbb{Z}^d$. In most, but not all, of the models we will consider, the couplings $J^\Lambda_{\{x,y\}}$ do not depend on Λ; in those cases, the set of all limits have an intrinsic definition as objects on all of $\mathbb{Z}^d$, which we will discuss later.

The simplest Ising model is the homogeneous one, where $J^\Lambda_{\{x,y\}}$ equals a constant $\check{J}$ for all Λ and all $\{x,y\} \in \mathbb{E}^d$. A model is called ferromagnetic (resp. antiferromagnetic) when $J^\Lambda_{\{x,y\}} \geq 0$ (resp. $J^\Lambda_{\{x,y\}} \leq 0$). When $J^\Lambda_{\{x,y\}} > 0$ and Λ is connected (by nearest neighbor bonds), it is clear that the free boundary condition ground states on Λ are $s_x \equiv +1$ and $s_x \equiv -1$, since these minimize each individual term in (0.2). The two constant configurations on all of $\mathbb{Z}^d$ are of course thus infinite volume ground states.

One of the basic results about the homogeneous ferromagnet is that, for $d \geq 2$ and large β ($\beta > \beta_c(d)$), there are distinct infinite volume Gibbs states corresponding to these two ground states, i.e., the limits of $P^+_{\Lambda,\beta}$ (with b.c. $\bar{s} \equiv +1$ on $\partial\Lambda$) and $P^-_{\Lambda,\beta}$ (with b.c. $\bar{s} \equiv -1$ on $\partial\Lambda$) are distinct. This is not so for small β ($\beta < \beta_c(d)$) or for $d = 1$. For the homogeneous ferromagnet, as we shall shortly see, there are also nonconstant infinite volume ground states. It is also known (for $d \geq 3$ and large β) that there are other infinite volume Gibbs states correponding to (some of) these ground states.

These lecture notes are concerned with disordered Ising models, in which the $J^\Lambda_{\{x,y\}}$'s are themselves random variables on some probability space $(\Omega, \mathcal{F}, \nu)$. Here the nature of infinite volume ground states and Gibbs states is more complicated and more interesting.

The model which is the most interesting and least understood is the Edwards-Anderson (EA) spin glass [EA75]. In this model $J^\Lambda_{\{x,y\}} = J_{\{x,y\}}(\omega)$, where $\{J_e : e \in \mathbb{E}^d\}$ is an i.i.d. family of symmetric (J_e equidistributed with $-J_e$) random variables, e.g., $\pm\check{J}$ valued or with a mean zero normal distribution. In the last part of these notes we will discuss various conjectures and controversies about the large β Gibbs states for the EA model. But before that we will treat three simpler situations. These lectures are thus divided into four chapters:

1. **Disordered ferromagnets.** Here $J^\Lambda_{\{x,y\}} = J_{\{x,y\}}(\omega)$, where the J_e's are i.i.d., but with $J_e \geq 0$. We will consider mainly $d = 2$, where the existence of nonconstant infinite volume ground states is equivalent to the existence of doubly infinite geodesics in a related first-passage percolation model. We will present a theorem [LiN96] supporting the conjecture that, for small d, nonconstant ground states do not exist (for ν-a.e. ω).

2. **Highly disordered systems.** We will consider a model [NS94], [NS96a], which for each Λ has, like the EA model, symmetric i.i.d. J^Λ_e's, but the magnitudes $|J^\Lambda_e|$ depend strongly on Λ. The issue of whether there exist multiple pairs of ground states for this model turns out to be equivalent to an interesting question about invasion percolation – whether there exist multiple trees in

the "invasion forest". Uniqueness of the tree is known for $d = 2$ [ChCN85] but there are only partial results for $d \geq 3$.

3. **Disordered systems at high temperature.** We will use the Fortuin-Kasteleyn (FK) percolation representation of the EA model to prove uniqueness of the infinite volume Gibbs distribution at high temperature [N94]. E.g., in the $\pm \breve{J}$ spin glass, this will be proved above the critical temperature of the homogeneous ferrromagnet.

4. **Disordered systems at low temperature.** We will discuss various concepts and conjectures from the physics literature on spin glasses, such as non-self-averaging, replica symmetry breaking, etc. The main focus will be on providing a coherent mathematical framework for these concepts and conjectures [NS92], [NS96b], [NS96c], [NS96d], [NS97].

Many important topics in the area of disordered Ising models, such as neural networks, are not considered at all in these lecture notes. We refer the reader to the papers in [BoP97] for reviews of some of these topics.

Chapter 1

Ground States of Disordered Ferromagnets

When $J_e^\Lambda = J_e$ (i.e., with no dependence on Λ), an infinite volume ground state for the couplings $\{J_e : e \in \mathbb{E}^d\}$ may be defined as any $s \in \mathcal{S} = \{-1, +1\}^{\mathbb{Z}^d}$, such that there exists some sequence $\bar{s}^{(L)} \in \mathcal{S}_{\partial \Lambda_L}$ (the restriction to cubes Λ_L and to fixed b.c.'s involves no loss of generality) and some ground state $s^{(L)} \in \mathcal{S}_{\Lambda_L}$ for $H_{\Lambda_L}^{\bar{s}^{(L)}}$ such that for each $x \in \mathbb{Z}^d$, $s_x = s_x^{(L)}$ for all large L. But there is an equivalent intrinsic characterization as follows.

For A a *finite* subset of $\mathbb{Z}^d$ and $s \in \mathcal{S}$, define $\Delta H_A(s)$ as

$$\Delta H_A(s) = 2 \sum_{\substack{x \in A, y \in \partial A \\ \{x,y\} \in \mathbb{E}^d}} J_{\{x,y\}} s_x s_y. \tag{1.1}$$

Formally, this is the change in total energy, i.e., in the formal Hamiltonian $H(s) = \sum_{x,y \in \mathbb{E}^d} -J_{\{x,y\}} s_x s_y$, when s is flipped (i.e., multiplied by -1) on A. We leave it as an exercise to show that s is an infinite volume ground state if and only if

$$\text{for all finite } A \subset \mathbb{Z}^d, \quad \Delta H_A(s) \geq 0. \tag{1.2}$$

Before considering disordered ferromagnets, we discuss infinite volume ground states for the homogeneous ferromagnet. For simplicity, set $d = 2$. In addition to the $s \equiv +1$ and $s \equiv -1$ constant ground states, one has also, for each integer j, a ground state s, where $s_x = +1$ for $x = (x_1, x_2)$ with $x_2 \geq j$ and $s_x = -1$ for $x_2 < j$. This is the limit, as $L \to \infty$, of the ground state on Λ_L with "Dobrushin boundary conditions", i.e., where $\bar{s}_x$ equals $+1$ for $x_2 \geq j$ and $\bar{s}_x$ equals -1 for $x_2 < j$.

In addition to these ground states, where the "interface" between $+$ and $-$ spins is a horizontal straight line and the analogous ones with vertical interfaces, any monotonic interface yields a ground state. In those cases, (1.2) is satisfied, although for some A's, $\Delta H_A(s) = 0$. Since one can choose an interface with any asymptotic angle θ, there are uncountably many infinite volume nonconstant ground states already in the homogeneous model.

5

For disordered ferromagnets, we will restrict attention to $\{J_e : e \in \mathbb{Z}^d\}$ that are i.i.d. with a common continuous distribution μ on $(0, \infty)$. One consequence of the continuity restriction is, that (a.s.) $\Delta H_A(s) \neq 0$ for all $s \in \mathcal{S}$ and all finite A. We begin with a proposition, which shows that in determining the existence of nonconstant ground states, it suffices to consider the ones with single interfaces. For $s \in \mathcal{S}$, we call $B \subset \mathbb{Z}^d$ a spin cluster for s, if (i) s_x is constant on B, (ii) B is connected (by nearest neighbor bonds) and (iii) there is no B' strictly containing B that satisfies (i) and (ii). Interfaces are the "surfaces" (in the dual lattice) separating spin clusters.

Proposition 1.1 *If $J_e > 0$ for every $e \in \mathbb{E}^d$ and there exists a nonconstant infinite volume ground state s for the J_e's, then there exists a nonconstant infinite volume ground state s^*, which has exactly two spin clusters (one $+$, one $-$) with both infinite.*

Proof. For ferromagnetic couplings, (1.1)-(1.2) immediately imply that if s is an infinite volume ground state and s^* is such that

$$\left\{ \{x, y\} \in \mathbb{E}^d : s_x^* \neq s_y^* \right\} \subseteq \left\{ \{x, y\} \in \mathbb{E}^d : s_x \neq s_y \right\}, \tag{1.3}$$

then s^* is also an infinite volume ground state. If s is nonconstant, then there exist $w, w' \in \mathbb{Z}^d$ such that $s_w \neq s_{w'}$. Let C' be the spin cluster for s which contains w' and define

$$\begin{aligned} C = \{x \in \mathbb{Z}^d : \quad &\text{there is a (nearest neighbor) path from} \\ &x \text{ to } w \text{ which avoids } C'\}; \end{aligned} \tag{1.4}$$

C is the connected component of w in the complement of C'. Note that from any point y in $\mathbb{Z}^d \setminus C$, there must be a path to C' (and hence to w') which avoids C; thus $\mathbb{Z}^d \setminus C$, like C, is connected. Define s^* to be $+1$ on C and -1 on $\mathbb{Z}^d \setminus C$ (so in particular $s_w^* = +1$ and $s_{w'}^* = -1$ and s^* is nonconstant). It is easy to see that C (and hence $\mathbb{Z}^d \setminus C$) is a union of spin clusters of s and thus (1.3) is valid. Thus s^* is an infinite volume ground state with exactly two spin clusters, C and $\mathbb{Z}^d \setminus C$. These are both infinite because each spin cluster B of the original s must be infinite; if it were finite, we could in (1.1) take $A = B$ making each term in the sum strictly negative (here we use $J_e > 0$) and thus violate (1.2). This completes the proof. $\qquad \square$

Let us now restrict attention to $d = 2$ and see what minimizing property is enjoyed by the interface between the two spin clusters of the s^* in Proposition 1.1. The interface is a doubly infinite path on the dual lattice $\mathbb{Z}^{2*} = \mathbb{Z}^2 + (1/2, 1/2)$. For each edge $e = \{x, y\} \in \mathbb{E}^d$, we denote by $e^* = \{x^*, y^*\}$ the dual nearest neighbor edge in $\mathbb{E}^{d*}$ (such that the line segment between x^* and y^* is the perpendicular bisector of the segment between x and y). Let us define $\tau(e^*) = 2J_e$ for such a dual pair of edges. For a finite (nearest neighbor path) r on $\mathbb{Z}^{2*}$, let us define

$$T(r) = \sum_{e^* \in r} \tau(e^*). \tag{1.5}$$

We can now characterize the interface of s^*.

Proposition 1.2 *Suppose $d = 2$, $J_e \geq 0$ for every $e \in \mathbb{E}^2$ and $s^* \in \mathcal{S}$ has exactly two spin clusters with both infinite. Then s^* is a ground state for $\{J_e : e \in \mathbb{E}^2\}$ if and only if its interface r^*, a doubly infinite path in $\mathbb{Z}^{2*}$, has the following property:*

For every x^*, y^* on r^*, the finite path segment $r^*[x^*, y^*]$ of r^* between x^* and y^* satisfies

$$T(r^*[x^*, y^*]) = \min\{T(r) : r \text{ is a path between } x^* \text{ and } y^*\}. \tag{1.6}$$

Proof. Suppose for some x^*, y^* on r^* and some path $\tilde{r}$ between x^* and y^*, $T(\tilde{r}) < T(r^*[x^*, y^*])$. By decomposing $\tilde{r}$ into portions along r^* and portions which are excursions from r^* (i.e., paths which only touch the sites of r^* at their start and finish), it is not hard to see that there is some x' and y' on r^* and an excursion r' from r^* which connects x' and y' such that $T(r') < T(r^*[x', y'])$. Thus the minimization in (1.6) may be restricted to paths r which are excursions from r^*. If s^* is a ground state and r' is any excursion from r^* which connects some x' and y', then the union of r' and $r^*[x', y']$ is a simple closed loop on $\mathbb{Z}^{2*}$ enclosing some finite subset A of $\mathbb{Z}^2$. Further,

$$\Delta H_A(s^*) = -T(r^*[x', y']) + T(r'). \tag{1.7}$$

Thus the ground state property (1.2) for s^* implies $T(r^*[x', y']) \leq T(r')$. Since this is true for every excursion r', we have (1.6) as desired.

The converse argument is as follows. Suppose (1.6) is valid. Let $A \subset \mathbb{Z}^d$ be finite. We want to show that $\Delta H_A = \Delta H_A(s^*) \geq 0$. Since ΔH_A is the sum of the ΔH_{A_j}'s for the connected components A_j of A, we may and will assume that A is connected. A may have one or more holes $B_1, B_2, \ldots$; these are the finite connected components of $\mathbb{Z}^2 \setminus A$. Each B_j has itself no holes because A is connected. Let $B_0 = A \cup B_1 \cup B_2 \cup \ldots$; this is A with holes filled in. Then the boundary of each B_j, for $j \geq 0$, is a simple closed loop (i.e., site self-avoiding) on $\mathbb{Z}^{2*}$ and ΔH_A is the sum of the ΔH_{B_j}'s, including $j = 0$. Thus we may and will assume that A has no holes. Let $\hat{r}$ denote the simple closed loop on $\mathbb{Z}^{2*}$, which is the boundary of such an A. We must show that (1.6) implies

$$\Delta H_A = -\sum_{e^* \in \hat{r} \cap r^*} \tau(e^*) + \sum_{e^* \in \hat{r} \setminus r^*} \tau(e^*) \geq 0. \tag{1.8}$$

This is trivially the case if $\hat{r} \cap r^*$ is empty. If it is nonempty, then $\hat{r}$ intersects at least two sites in r^* and we can write the second sum of (1.8) as a sum of $T(r_j)$ over the excursions of $\hat{r}$ from r^*. Then by (1.6)

$$\Delta H_A \geq -\sum_{e^* \in \hat{r} \cap r^*} \tau(e^*) + \sum_j T(r_j^*), \tag{1.9}$$

where r_j^* is the segment of r^* connecting the endpoints of the excursion r_j. Because $\hat{r}$ was a closed loop, $\tau(e^*)$ for each $e^* \in \hat{r} \cap r^*$ must appear in at least one of the $T(r_j^*)$'s and so the RHS of (1.9) is ≥ 0, which completes the proof. $\square$

The RHS of (1.6) is exactly the definition of the passage time $T(x^*, y^*)$ for the standard first-passage percolation model on $\mathbb{Z}^{2*}$ with i.i.d. non-negative $\tau(e^*)$'s [HaW65]. Since $\tau(e^*) > 0$ a.s., $T(x^*, y^*)$ is a (random) metric on $\mathbb{Z}^{2*}$ and the defining property of r^* given in Proposition 1.2 is exactly that of r^* being a doubly infinite geodesic for this metric (called a bigeodesic in [LiN96]). Propositions 1.1 and 1.2 together show that for $d = 2$, nonconstant ground states exist for a disordered ferromagnet if and only if bigeodesics exist for the associated first-passage percolation metric.

For $d \geq 3$, Proposition 1.1 remains valid, but Proposition 1.2 must be replaced by one involving "minimal surfaces" (of codimension one) formed by the elementary codimension one faces of the unit cubes of $\mathbb{Z}^{d*}$. Thus first-passage percolation is replaced by a model (see [Kes87]) assigning i.i.d. variables $\tau(e^*) = 2J_e$ to the faces e^* dual to the edges e of $\mathbb{E}^d$. Paths r' connecting x' and y' are replaced by surfaces formed out of these faces with a specified (codimension two) boundary and $T(r')$ is replaced by the sum of $\tau(e^*)$ over the faces e^* forming the surface. (We remark that somewhat differently defined minimal surfaces are studied in [Bo96].) Bigeodesics are replaced by infinite surfaces with no boundary such that each finite portion of the surface minimizes this sum with the boundary of that portion fixed. We note that such minimal surfaces and the related ground states are studied in [W96].

According to nonrigorous scaling arguments in the physics literature (see, e.g., Section 5 of [FoLN91]), nonconstant ground states for disordered ferromagnets should not exist (presumably in the a.s. sense) for low d (including $d = 2$), but should exist for large d. For $d = 1$, with a continuous common distribution for the J_e's, it is an elementary exercise (which we recommend to the reader) to show that a.s. there are no nonconstant ground states. For no other d is there a complete result. (But see [BoK94] and [BoK96] for rigorous results in high and low dimension about a different class of interface models.) The remainder of this chapter concerns results for $d = 2$, which support the a.s. nonexistence conjecture (see also [WW96]). Other results, valid for all $d \geq 2$, have been obtained by Wehr [W96], who showed that with a common continuous distribution of finite mean for the J_e's, the number of nonconstant ground states is either a.s. zero or else a.s. infinite.

Our results concern bigeodesics with specified asymptotic directions, defined as follows. We define a unigeodesic as a semi-infinite path on $\mathbb{Z}^{2*}$ satisfying (1.6) for every x^*, y^* on the path. Thus a bigeodesic is always a union of two unigeodesics (but the converse is not generally valid). If $\hat{x}$ is a unit vector (a vector in $\mathbb{R}^2$ of Euclidean length one) and r is a unigeodesic (starting at $x_0^* \in \mathbb{Z}^{2*}$ and consisting of the sites $x_0^*, x_1^*, \ldots$ and the edges $\{x_0^*, x_1^*\}, \ldots$) we say that r is an $\hat{x}$-unigeodesic if $x_n^*/\|x_n^*\| \to \hat{x}$ as $n \to \infty$. A bigeodesic is said to be an $(\hat{x}, \hat{y})$-bigeodesic if it is the union of an $\hat{x}$-unigeodesic and a $\hat{y}$-unigeodesic.

The heuristic scaling argument against the existence of bigeodesics is really an argument against $(\hat{x}, -\hat{x})$-bigeodesics for any deterministic $\hat{x}$, which proceeds roughly as follows. Let $y_n^* \in \mathbb{Z}^{2*}$ be a deterministic sequence with $\|y_n^*\| \to \infty$ and

$y_n^*/||y_n^*|| \to \hat{x}$ and consider the finite geodesic M_n connecting y_n^* and $-y_n^*$. If there were an $(\hat{x}, -\hat{x})$-bigeodesic, M_n should converge to it. But the closest distance of M_n to the origin should scale like $||y_n^*||^\xi$ with $\xi > 0$, and so M_n should have no limit. The next theorem rules out $(\hat{x}, -\hat{x})$-bigeodesics with a deterministic $\hat{x}$, at least for most $\hat{x}$'s. Essentially the same arguments rule out $(\hat{x}, \hat{y})$-bigeodesics (see [LiN96]). In Appendix A, we explain why various sets in Ω, the underlying probability space of the $\tau(e^*)$'s, considered in this chapter, such as $\{\exists$ an $(\hat{x}, \hat{y})$-bigeodesic$\}$, are measurable. I.e., we show that such a set differs from an event in the σ-field generated by the $\tau(e^*)$'s by a subset of a zero-probability event. Following the theorem, we discuss the strong conjecture that a.s. no bigeodesics of any sort exist. Then we give the proof of the theorem (as in [LiN96] but with simplifications suggested by C.D. Howard). At the end of the chapter (see Theorem 1.5 and the discussion preceding it), we give an improvement of the next theorem, due to M.P.W. Zerner, in which "Lebesgue-a.e. $\hat{x}$" is replaced by "all but countably many $\hat{x}$'s".

Theorem 1.3 ([LiN96]) *Suppose $d = 2$ and the $\tau(e^*)$'s are continuously distributed, then for Lebesgue-a.e. $\hat{x}$, there is a.s. no $(\hat{x}, -\hat{x})$-bigeodesic.*

Remarks. For a class of first-passage percolation models based on the two-dimensional Poisson process (rather than on a deterministic lattice), an improved analogue of Theorem 1.3 has been obtained, where the result is valid for *every* (deterministic) direction $\hat{x}$ [HoN96]. The results and arguments of [N95] imply (under a certain curvature assumption on the first-passage asymptotic shape, which is reasonable but unfortunately has not been verified) that a.s. there are no bigeodesics other than $(\hat{x}, -\hat{x})$-ones. It is likely that in the Poisson process model of [HoN96], such a result will be proved without unverified hypothesis. The most serious gap between Theorem 1.3 and the strong conjecture that a.s. no geodesics exist at all, is to rule out $(\hat{x}, -\hat{x})$-bigeodesics with $\hat{x}$ depending on ω. One can do a more sophisticated scaling argument than the one discussed above, which suggests that this stronger conjecture should be valid providing one has superdiffusivity, in the sense that $\xi > 1/2$ (rather than just $\xi > 0$). ξ is conjectured to equal $2/3$ for $d = 2$ (see [HuH85], [K85], [HuHF85], [KPZ86]) and there is a recent rigorous lower bound of $3/5$ [LiNP96], but for a technically different definition of ξ. The strong conjecture has much support, but not as yet a proof.

Proof. Because the $\tau(e^*)$'s are continuously distributed, for each $u, v \in \mathbb{Z}^{2*}$, there is (a.s.) a unique finite geodesic, $M(u, v)$, connecting u and v (such that $T(u, v) = T(M(u, v))$). Define $R(u)$ to be the graph with vertex set $\mathbb{Z}^{2*}$ and edge set consisting of all edges in $\mathbb{E}^{2*}$ appearing in $M(u, v)$ for some v. By constructing $R(u)$ sequentially, adding one $M(u, v)$ at a time, it is not hard to see that $R(u)$ is a tree, which clearly spans all of $\mathbb{E}^{2*}$. Any infinite path in this tree starting at u is a unigeodesic. Define the event

$$D_U(\hat{x}) = \{ \text{ for every } u \text{ in } \mathbb{Z}^{2*}, \text{ there is at most one}$$
$$\hat{x}\text{-unigeodesic starting at } u\}, \qquad (1.10)$$

and define $\mathcal{U}$ to be the set of $\hat{x}$'s such that $D_U(\hat{x})$ a.s. occurs, i.e., $\nu(D_U(\hat{x})) = 1$. We will first show that $\mathcal{U}$ has full Lebesgue measure (this result about $\mathcal{U}$ will be improved in Theorem 1.5 below) and then go on to prove that for every $\hat{x}$ in $\mathcal{U}$, there is a.s. no $(\hat{x}, -\hat{x})$-bigeodesic.

Let us define $\hat{R}(u)$ to be the subgraph of $R(u)$, which contains only the sites and edges of $R(u)$ that belong to unigeodesics starting from u. ($\hat{R}(u)$ is nonempty, since $R(u)$, as an infinite tree, must contain at least one infinite path starting from u.) If the edge $\{u, v\}$ is in $\hat{R}(u)$, we now define a specific unigeodesic $\hat{r}^+(u, v)$, starting from u; if $\{u, v\}$ is not in $\hat{R}(u)$, $\hat{r}^+(u, v)$ will be undefined. $\hat{r}^+(u, v)$ is defined sequentially as the infinite path $u_0 = u$, $u_1 = v$, u_2, ..., with each $\{u_{j-1}, u_j\}$ in $\hat{R}(u)$, such that, given $(u_0, \ldots, u_j)$, u_{j+1} is chosen so that the angle from $u_j - u_{j-1}$ to $u_{j+1} - u_j$ (which can be $\pi/2$, 0 or $-\pi/2$) is the maximum possible. $\hat{r}^-(u, v)$ is defined similarly but with minimum possible angle.

Suppose there are two distinct $\hat{x}$-unigeodesics r_1 and r_2 starting from some x^*. They must bifurcate at some u going to v_1 and v_2 in their next steps. Because $d = 2$, any unigeodesic caught "between" r_1 and r_2 must also be an $\hat{x}$-unigeodesic; this will be so either for $\hat{r}^+(u, v_2)$ (and $\hat{r}^-(u, v_1)$) or else for $\hat{r}^+(u, v_1)$ (and $\hat{r}^-(u, v_2)$). Thus $D_U(\hat{x})$ occurs unless the event $G(u, v, \hat{x})$, that $\hat{r}^+(u, v)$ is an $\hat{x}$-geodesic occurs for some (u, v), and so

$$1 \geq \nu(D_U(\hat{x})) \geq 1 - \sum_{(u,v)} \nu(G(u, v, \hat{x})). \tag{1.11}$$

But for each (u, v), $\hat{r}^+(u, v)$ can be an $\hat{x}$-unigeodesic for at most one $\hat{x}$ and so by Fubini's theorem

$$\int \nu(G(u, v, \hat{x}))d\hat{x} = \int \left[\int 1_{G(u,v,\hat{x})}(\omega)d\hat{x}\right] \nu(d\omega) = 0, \tag{1.12}$$

where $d\hat{x}$ denotes uniform (i.e., normalized Lebesgue) measure. Integrating (1.11) against $d\hat{x}$ then implies that $\nu(D_U(\hat{x})) = 1$ for Lebesgue-a.e. $\hat{x}$; i.e., $\mathcal{U}$ has full Lebesgue measure.

In order to prove that for $\hat{x}$ in $\mathcal{U}$, there is a.s. no $(\hat{x}, -\hat{x})$-bigeodesic, we will show (see Proposition 1.4 below) that for $\hat{x}$ in $\mathcal{U}$ the (a.s. unique, if they exist) $\hat{x}$-unigeodesics starting from u and v must meet and hence coalesce (since $\hat{x}$ is in $\mathcal{U}$). Once we have done this, the completition of the proof is as follows.

Since $\hat{x} \in \mathcal{U}$ implies $-\hat{x} \in \mathcal{U}$ (by reflection symmetry), if there were two distinct $(\hat{x}, -\hat{x})$-bigeodesics, they would have to coalesce in both the $\hat{x}$ and $-\hat{x}$ directions yielding two distinct finite geodesics between some pair of sites, which would contradict the uniqueness of finite geodesics. Thus there is a.s. at most one $(\hat{x}, -\hat{x})$-bigeodesic. Let A denote the event that there is exactly one $(\hat{x}, -\hat{x})$-bigeodesic; we must show $\nu(A) = 0$. For $u \in \mathbb{Z}^{2*}$ and $L > 0$, let $A(L, u)$ denote the event that there is exactly one $(\hat{x}, -\hat{x})$-bigeodesic and it touches some site in $u + \{-L, \ldots, L\}^2$. If $\nu(A) > 0$, then for some large L, $\nu(A(L, 0)) > 0$ and so by translation invariance $\nu(A(L, z)) = \varepsilon > 0$ for every z. Let $z_n \in \mathbb{Z}^{2*}$ be such that

$||z_n|| \to \infty$, while $z_n/||z_n|| \to \hat{z} \neq \pm\hat{x}$. Then, on the event $A(L, z)$, the unique $(\hat{x}, -\hat{x})$-bigeodesic can touch $z_n + \{-L, \ldots, L\}^2$ for only finitely many n's and so for any z,

$$\nu(A(L, z) \cap A(L, z_n)) \to 0 \text{ as } n \to \infty. \tag{1.13}$$

Taking $z = z_m$, we see that for any $\delta > 0$, there is some (deterministic) subsequence $w_k = z_{n_k}$ such that $\nu(A(L, w_k) \cap A(L, w_j)) \leq \delta$ for all $j \neq k$. But then

$$\nu\left(\bigcup_{k=1}^{N} A(L, w_k)\right) \geq N\varepsilon - \binom{N}{2}\delta. \tag{1.14}$$

Choosing $N > 2/\varepsilon$ and then $\delta < \binom{N}{2}^{-1}$ gives a probability bigger than one, contradicting the assumption that $\nu(A) > 0$.

It remains only to prove that $\hat{x}$-unigeodesics starting from u and v must intersect (and so coalesce). We state this as a separate proposition. $\qquad\square$

Proposition 1.4 ([LiN96]) *Suppose that $d = 2$, that the $\tau(e^*)$'s are continuously distributed and that $\hat{x} \in \mathcal{U}$; then there is zero probability that there exist disjoint $\hat{x}$-unigeodesics.*

Proof. Let $r^{\#}(u)$ denote the (unique, if it exists) $\hat{x}$-unigeodesic starting from u. Let $\mathcal{R}$ denote the graph with vertex set $\{u \in \mathbb{Z}^{2*} : r^{\#}(u) \text{ exists }\}$ and edge set $\{e^* \in \mathbb{E}^{2*} : e^* \text{ belongs to some } r^{\#}(u)\}$. Since $r^{\#}(u)$ and $r^{\#}(v)$ must coalesce if they ever meet, it follows that $\mathcal{R}$ is either empty or else is a forest consisting of $N \geq 1$ trees, all infinite. Proposition 1.4 is equivalent to the statement, that $\nu(N \geq 2) = 0$. For the remainder of the proof (in particular for Part 2 below), we will assume (without loss of generality), that $(\Omega, \mathcal{F}, \nu)$ is the product probability space (over $\mathbb{E}^{2*}$) of $(\mathbb{R}, \mathcal{B}, \mu^*)$, where $\mathcal{B}$ is the Borel σ-field and μ^* is the common distribution of the $\tau(e^*)$'s. In particular $\tau(e^*)$ is the coordinate function ω_{e^*}. The structure of the proof is parallel to that used in [BuK89] to show uniqueness of infinite percolation clusters; it has three parts.

Part 1. $\nu(N \geq 2) > 0$ *implies* $\nu(N \geq 3) > 0$. Without loss of generality, we assume $\hat{x} = (\hat{x}_1, \hat{x}_2)$ with $\hat{x}_1 > 0$. By translation invariance and the fact that any $\hat{x}$-unigeodesic is eventually to the right of any vertical line, $\nu(N \geq 2) > 0$ implies, that for some integers $j_1 < j_2$, $\nu(A(j_1, j_2)) > 0$, where, for integers $j_1 < j_2 < \ldots < j_m$, $A(j_1, \ldots, j_m)$ denotes the event, that $r_1^{\#} = r^{\#}((-1/2, 1/2 + j_1)), \ldots, r_m^{\#} = r^{\#}((-1/2, 1/2 + j_m))$ exist, are disjoint and touch only sites (after the initial one) with positive first coordinate. Since $\nu(A(j_1 + m, j_2 + m))$ is strictly positive and does not depend on m, it follows that $\nu(A(j_1, j_2) \cap A(j_1 + m, j_2 + m)) \not\to 0$ as $m \to \infty$ and hence, by choosing m large, $\nu(A(j_1, j_2) \cap A(j_2', j_3)) > 0$ for some $j_2' > j_2$ (with $j_3 = j_2' + j_2 - j_1$). On this last event, even though the two middle geodesics, $r_2^{\#}$ and $r_2^{\#\prime}$, may coalesce, $r_2^{\#}$ cannot meet $r_3^{\#}$, since $r_2^{\#\prime}$ is caught between them and it cannot meet $r_3^\#$. This last event is thus contained in the event $A(j_1, j_2, j_3)$ and so

$$\nu(A(j_1, j_2, j_3)) > 0. \tag{1.15}$$

This positivity for some $j_1 < j_2 < j_3$ implies, and is in fact equivalent to, $\nu(N \geq 3) > 0$, which completes the proof of Part 1. We note that similar arguments show that $\nu(N \geq 2) > 0$ is equivalent to $\nu(N = \infty) = 1$, but we will not need this.

Part 2. $\nu(N \geq 3) > 0$ *implies* $\nu(\hat{F}_{m,k}) > 0$ *for some* $m, k \geq 1$. Here $\hat{F}_{m,k}$ for positive integers m and k denotes the event that some tree in $\mathcal{R}$ touches the rectangle $\hat{R}_{m,k} = \{u = (\alpha, \beta) \in \mathbb{Z}^{2*} : -m < \alpha < 0, \ 0 < \beta < k\}$ but no other sites in the left half plane. The proof of this part includes some simplifications from that given in [LiN96] suggested by C.D. Howard.

We start from (1.15), where by translation invariance we may assume $j_1 = -1$. We will then take $k = j_3$ and determine m later, although we remark now that m=1 will suffice if the support of μ^* is unbounded. The strategy, roughly speaking, is to take some ω's from $A(j_1, j_2, j_3)$ and increase the $\tau(e^*)$'s for edges inside $\hat{R}_{m,k}$ enough to create ω''s in $\hat{F}_{m,k}$.

Define $\hat{\Theta} = \hat{\Theta}_{m,k}$ for $m = 1$ as the set of edges $\{x^*, y^*\}$ with $y^* \in \hat{R}_{1,k}$ and x^* the western neighbor of y^*, and for $m > 1$ as the set of edges $\{x^*, y^*\} \in \mathbb{E}^{2*}$ with $x^* \in \hat{R}_{m,k} \setminus \hat{R}_{1,k}$. Define for $m > 1$, $\partial^a = \partial^a_{m,k}$ for $a = W$ or N or S to be the western or northern or southern boundary of $\hat{R}_{m,k} \setminus \hat{R}_{1,k}$, while for $m = 1$, $\partial^W_{1,k}$ is the western boundary of $\hat{R}_{1,k}$ and $\partial^a_{1,k}$ is empty for $a = N$ or S. For example, $\partial^W_{m,k} = \hat{R}_{1,k} - m$. The role of the eastern boundary will be played by $\partial^E = \partial^E_{m,k} = \hat{R}_{1,k}$. We also need corresponding sets of edges. For $a = W$ or E this is,

$$\tilde{\partial}^a = \tilde{\partial}^a_{m,k} = \{ \text{ vertical edges } \{x^*, y^*\} \in \mathbb{E}^{2*} : x^* \in \partial^a_{m,k} \}; \tag{1.16}$$

for $a = N$ or S, vertical is simply replaced by horizontal for $m > 1$, while for $m = 1$, $\tilde{\partial}^a_{1,k}$ is the single edge going west from $(-1/2, k+1/2)$ (resp., from $(-1/2, -1/2)$) for $a = N$ (resp., for $a = S$).

For our fixed μ^*, $j_1 = -1 < j_2 < j_3$ and $\varepsilon = \nu(A(j_1, j_2, j_3)) > 0$, we choose λ sufficiently large so that the event

$$\hat{B}^\lambda = \hat{B}^\lambda_{m,k} = \left\{ \sum_{e^* \in \tilde{\partial}^W} \tau(e^*) + \sum_{e^* \in \tilde{\partial}^E} \tau(e^*) < \lambda \right\} \tag{1.17}$$

has $\nu(\hat{B}^\lambda) \geq 1 - \varepsilon/5$. The probability $\nu(\hat{B}^\lambda_{m,k})$ is easily seen to depend only on μ^* and k but not on m, so that the same λ works for all m. We are now ready to pick m. Let $\bar{\lambda}$ denote the supremum of the support of μ^*; it may be $+\infty$. For any $\lambda' < \bar{\lambda}$, we consider for $a = N$ or S, the event

$$\hat{B}^{a,\lambda,\lambda'} = \hat{B}^{a,\lambda,\lambda'}_{m,k} = \left\{ \sum_{e^* \in \tilde{\partial}^a} \tau(e^*) < m\lambda' - \lambda \right\}. \tag{1.18}$$

If $\bar{\lambda} = \infty$, we choose $m = 1$ and then λ' so large that for $a = N$ or S,

$$\nu\left(\hat{B}_{1,k}^{a,\lambda,\lambda'}\right) = \nu\left(\tau(e^*) < \lambda' - \lambda\right) \geq 1 - \varepsilon/5. \tag{1.19}$$

If $\bar{\lambda} < \infty$, we first choose λ' strictly between $E(\tau(e^*))$ and $\bar{\lambda}$, so that (for $a = N$ or S and fixed k), by the law of large numbers,

$$\lim_{m \to \infty} m^{-1} \sum_{e^* \in \tilde{\partial}_{m,k}^a} \tau(e^*) = E(\tau(e^*)) < \lambda'. \tag{1.20}$$

This implies that we may choose m sufficiently large so that again $\nu\left(\hat{B}_{m,k}^{a,\lambda,\lambda'}\right) \geq 1 - \varepsilon/5$ for $a = N$ or S. For our chosen m and k, we now take $\lambda'' < \bar{\lambda}$ sufficiently large so that the event

$$\tilde{B}^{\lambda''} = \tilde{B}_{m,k}^{\lambda''} = \left\{\tau(e^*) < \lambda'' \text{ for every } e^* \in \tilde{\partial}^N \cup \tilde{\partial}^S \cup \tilde{\partial}^E\right\} \tag{1.21}$$

also has $\nu(\tilde{B}) \geq 1 - \varepsilon/5$.

Let $\hat{A}$ denote the intersection

$$\hat{A} = A(j_1, j_2, j_3) \cap \hat{B}^\lambda \cap \hat{B}^{N,\lambda,\lambda'} \cap \hat{B}^{S,\lambda,\lambda'} \cap \tilde{B}^{\lambda''}, \tag{1.22}$$

where k, λ, λ' and m have been chosen as indicated above. Then $\nu(\hat{A}^c) \leq 1 - \varepsilon + 4(\varepsilon/5)$, so $\nu(\hat{A}) \geq \varepsilon/5 > 0$. For $\omega \in \hat{A}$, any $u \in \partial^W$ and $v \in \partial^E$, there is a path r between u and v using only the edges of $\partial^W \cup \partial^N \cup \partial^E$ with $T(r) < m\lambda'$. For $\omega \in \hat{A}$, any $u = (-1/2 - i, 1/2 + k) \in \partial^N$ (resp., $u = (-1/2 - i, -1/2) \in \partial^S$) and $v = (-1/2, 1/2 + j) \in \partial^E$, there is a path r between u and v using only the edges of $\partial^N \cup \partial^E$ (resp., $\partial^S \cup \partial^E$) with $T(r) < (i + k - j)\lambda''$ (resp., $T(r) < (i + j + 1)\lambda'')$.

Let $\tilde{\lambda}$ be some number, to be chosen below, satisfying $\max(\lambda', \lambda'') \leq \tilde{\lambda} < \bar{\lambda}$. Suppose $\omega \in \hat{A}$ is altered to ω' by changing each $\omega_{e^*} < \tilde{\lambda} \equiv \max(\lambda', \lambda'')$ with $e^* \in \hat{\Theta}$ to some value at least as large as $\tilde{\lambda}$. Since the ω_{e^*}'s with e^* in any of the three unigeodesics $r_i^\#$ are unchanged while others increase or stay constant, it follows that each $r_i^\#$ is still an $\hat{x}$-unigeodesic for ω'. Thus ω' is still in $A(j_1, j_2, j_3)$ and in fact is still in $\hat{A}$. On the other hand, for ω', any finite geodesic $M(u, v)$ with $u \in \partial^W$ and $v \in \partial^E$ cannot use only edges in $\hat{\Theta}$, since such a path r' would have $T(r') \geq m\tilde{\lambda}$, while there is a path r not using any edges in $\hat{\Theta}$ with $T(r) < m\lambda' \leq m\tilde{\lambda}$. Similarly, any $M(u, v)$ with $u \in \partial^N \cup \partial^S$ and $v \in \partial^E$ cannot use only edges in $\hat{\Theta}$, since there would be a better path r not using edges in $\hat{\Theta}$. It follows that any geodesic $\tilde{r}$ for ω' which starts from $u = (\alpha, \beta)$ with $\alpha < 0$ and $(\alpha, \beta) \notin \hat{R}_{m,k}$ cannot touch the middle geodesic $r_2^\#$ without first passing through either $r_1^\#$ or $r_3^\#$. If this $\tilde{r}$ were an $\hat{x}$-unigeodesic, it could only touch $r_2^\#$ if ω' had the property that some $\hat{x}$-unigeodesics meet without coalescing. Thus any event A^* consisting of such ω''s would have $A^* \cap D_U(\hat{x}) \subseteq \hat{F}_{m,k}$. Since $\hat{x} \in \mathcal{U}$, we thus would have $\nu(\hat{F}_{m,k}) \geq \nu(A^*)$. To complete the proof of Part 2, it remains only to choose $\tilde{\lambda}$ and construct such an A^* with $\nu(A^*) > 0$.

We choose $\tilde{\lambda}$ sufficiently close to $\bar{\lambda}$ so that the event

$$\tilde{A} = \hat{A} \cap \{\tau(e^*) < \tilde{\lambda} \text{ for every } e^* \in \hat{\Theta}\}, \qquad (1.23)$$

which converges to $\hat{A}$ as $\tilde{\lambda} \to \bar{\lambda}$, has $\nu(\tilde{A}) \geq \varepsilon/10 > 0$. Our desired A^* is constructed from $\tilde{A}$ as follows. With $\Omega = \mathbb{R}^{\mathbb{E}^{2*}}$, we define $\Omega_1 = \mathbb{R}^{\mathbb{E}^{2*}\backslash\hat{\Theta}}$ and $\Omega_2 = \mathbb{R}^{\hat{\Theta}}$ and write $\omega \in \Omega$ as (ω_1, ω_2) with ω_i in Ω_i. We also have $\nu = \nu_1 \times \nu_2$ with ν_i the obvious product measure on Ω_i. For $\omega_1 \in \Omega_1$, we define the Ω_2-section of $\tilde{A}$ as

$$\tilde{A}_{\omega_1} = \{\omega_2 \in \Omega_2 : (\omega_1, \omega_2) \in \tilde{A}\}. \qquad (1.24)$$

Then

$$\nu(\tilde{A}) = \int_{\Omega_1} \nu_2(\tilde{A}_{\omega_1})\nu_1(d\omega_1) > 0, \qquad (1.25)$$

so

$$\nu_1\left(\{\omega_1 \in \Omega_1 : \nu_2(\tilde{A}_{\omega_1}) > 0\}\right) > 0. \qquad (1.26)$$

Finally, we define

$$A^* = \left\{\omega_1 \in \Omega_1 : \nu_2(\tilde{A}_{\omega_1}) > 0\right\} \times [\tilde{\lambda}, \bar{\lambda})^{\hat{\Theta}}. \qquad (1.27)$$

It should be clear that (a) $A^* \in \mathcal{F} = \mathcal{B}^{\mathbb{E}^{2*}}$; (b) every $\omega' \in A^*$ may be regarded as an altered ω from $\tilde{A} \subset \hat{A}$, as described above; and (c) $\nu(A^*) > 0$ because of (1.26) and the product structure of ν.

Part 3. $\nu(\hat{F}_{m,k}) > 0$ *leads to a contradiction.* Consider a rectangular array of nonintersecting translates $\hat{R}^x_{m,k}$ of the basic rectangle $\hat{R}_{m,k}$ indexed by $x \in \mathbb{Z}^2$ (in a natural way) and the corresponding translated events $\hat{F}^x_{m,k}$. If both $\hat{F}^x_{m,k}$ and $\hat{F}^y_{m,k}$ occur with $x \neq y$, then, from the definition of these two events, the corresponding trees in $\mathcal{R}$ must be disjoint. Let n_L denote the number of $\hat{R}^x_{m,k}$'s inside $[0, L] \times [0, L]$ and N_L the random number of corresponding $\hat{F}^x_{m,k}$'s which occur. Translation invariance implies $E(N_L) = n_L \nu(\hat{F}_{m,k})$. Since $n_L \geq cL^2$ for some $c > 0$ and all large L, we see that $\nu(\hat{F}_{m,k}) > 0$ would imply

$$\nu(N_L \geq c'L^2) > 0 \text{ for all large L}, \qquad (1.28)$$

with $c' = c\nu(\hat{F}_{m,k}) > 0$. But since all N_L trees are disjoint and they are all infinite, they must touch at least N_L distinct points on the boundary of $\mathbb{Z}^{2*} \cap ([0, L] \times [0, L])$. But the number of boundary sites is bounded by $c''L$ for large L with $c'' < \infty$, which leads to the contradiction that $c'L^2 \leq c''L$ for all large L. This proves Part 3.

Combining Parts 1, 2 and 3, we conclude that $\nu(N \geq 2) = 0$, which, as noted earlier, is equivalent to the claim of Proposition 1.4, whose proof is now complete. $\qquad \square$

We have shown that for any $\hat{x}$ in $\mathcal{U}$, there is a.s. no $(\hat{x}, -\hat{x})$-bigeodesic. It is natural to conjecture that all unit vectors $\hat{x}$ are in $\mathcal{U}$. The next theorem is weaker than that but stronger than the result above that $\mathcal{U}$ has full Lebesgue measure.

Theorem 1.5 ([Z96]) *Suppose $d = 2$ and the $\tau(e^*)$'s are continuously distributed. Then $\mathcal{U}$ contains all but (at most) countably many unit vectors $\hat{x}$.*

Proof. Let $C_u(\hat{x})$ denote the event that there exist $\hat{x}$-unigeodesics r_1 and r_2, with $r_1 = (u, u_1^1, u_2^1, \ldots)$ and $r_2 = (u, u_1^2, u_2^2, \ldots)$ and $u_1^1 \neq u_1^2$. Then, by (A.9) in Appendix A and translation invariance,

$$\hat{x} \notin \mathcal{U} \iff \nu \left(\bigcup_{u \in \mathbb{Z}^{2*}} C_u(\hat{x}) \right) > 0 \iff \nu\left(C(\hat{x})\right) > 0, \qquad (1.29)$$

where $C(\hat{x})$ denotes $C_{u'}(\hat{x})$ for some fixed $u' \in \mathbb{Z}^{2*}$, e.g., $u' = (1/2, 1/2)$. We will next assume the contrary of what is to be proved and show that this leads to a contradiction. If $\nu\left(C(\hat{x})\right) > 0$ for uncountably many $\hat{x}$'s, then for some positive integer m, there are uncountably many $\hat{x}$'s with $\nu\left(C(\hat{x})\right) \geq 1/m$. Then, taking $n > 4m$, we would have distinct $\hat{x}_1, \ldots, \hat{x}_n$ such that

$$\sum_{i=1}^{n} \nu\left(C(\hat{x}_i)\right) = E\left(\sum_{i=1}^{n} 1_{C(\hat{x}_i)} \right) > 4, \qquad (1.30)$$

where E denotes expectation w.r.t. ν. Then the event that 5 or more of the $C(\hat{x}_i)$'s occur simultaneously has positive probability and so there are 5 distinct unit vectors $\hat{y}_1, \ldots, \hat{y}_5$ (taken from $\{\hat{x}_1, \ldots, \hat{x}_n\}$) such that

$$\nu\left(C(\hat{y}_1) \cap \ldots \cap C(\hat{y}_5)\right) > 0. \qquad (1.31)$$

On the intersection event in (1.31), there are ten different unigeodesics starting from u' and so among the four nearest neighbors of u' there must be (at least) one, v', which is the second point of at least three unigeodesics, corresponding to three distinct $\hat{y}_i$'s. Thus for some v' and some three distinct unit vectors $\hat{z}_1, \hat{z}_2, \hat{z}_3$ (taken from $\{\hat{y}_1, \ldots, \hat{y}_5\}$),

$$\nu\left(\exists \text{ for } i = 1, 2, 3, \text{ a } \hat{z}_i\text{-unigeodesic } r_i \text{ starting with } (u', v')\right) > 0. \qquad (1.32)$$

By choice of labelling, we may assume that when the r_i's bifurcate from each other, r_1 is the most clockwise, r_3 is the most counterclockwise and r_2 is in the middle. Since finite geodesics between pairs of sites are a.s. unique, none of the three "partner" unigeodesics $\bar{r}_i$ (which are also $\hat{z}_i$-unigeodesics but starting with (u', v_i'') with $v_i'' \neq v'$) can touch any site (other than u') in r_1 or r_2 or r_3. In particular, $\bar{r}_2$ cannot touch r_1 or r_3. But since $\hat{z}_1$ and $\hat{z}_3$ are distinct from $\hat{z}_2$, this makes it impossible for $\bar{r}_2$ to be a $\hat{z}_2$-unigeodesic. This contradiction proves that $\nu\left(C(\hat{x})\right) = 0$ for all but countably many $\hat{x}$'s, which completes the proof. $\square$

Chapter 2

Ground States of Highly Disordered Systems

In this chapter, based on [NS94] and [NS96a] (see also [CiMB94]), we consider, for each cube Λ_L, couplings $J_e^{(L)} = J_e^{\Lambda_L}$ which, like in the EA spin glass model, are i.i.d. symmetric random variables, on a fixed probability space $(\Omega, \mathcal{F}, \nu)$. However, unlike the EA model, these couplings will (strongly) depend on L. The dependence, explained precisely below, is in terms of a fixed set of random variables $\{K_e, \alpha_e : e \in \mathbb{E}^d\}$ and a nonlinear scaling parameter $\lambda^{(L)}$. The sign of $J_e^{(L)}$ will simply be α_e for every L while the magnitude $|J_e^{(L)}|$ will be determined by K_e and $\lambda^{(L)}$. Infinite volume ground states, defined as the limits of finite volume ground states $s^{(L)}$ for $H_{\Lambda_L}^{\bar{s}^{(L)}}$, with arbitrary boundary conditions $\bar{s}^{(L)}$, will again have an intrinsic characterization. But here the characterization, in terms of $\{K_e, \alpha_e : e \in \mathbb{E}^d\}$, will take some time to derive. Since flipping the b.c. $\bar{s}^{(L)}$ to $-\bar{s}^{(L)}$ clearly flips the ground state $s^{(L)}$ to $-s^{(L)}$, it follows that infinite volume ground states always come in pairs. The main issue we will address is whether or not there is a.s. only a single pair.

The dependence of $J_e^{(L)}$ on L is chosen so that each of the $J_e^{(L)}$'s appearing in $H_{\Lambda_L}^{\bar{s}^{(L)}}$ has a magnitude "on its own scale". More precisely, let m_L denote the number of edges $e = \{x, y\} \in \mathbb{E}^d$ with x or y (or both) in Λ_L and arrange the m_L random variables $|J_e^{(L)}|$, for these e's, in rank order,

$$|J_{(1)}^{(L)}| \geq |J_{(2)}^{(L)}| \geq \ldots \geq |J_{(m_L)}^{(L)}|; \tag{2.1}$$

our definition of the $J_e^{(L)}$'s will guarantee that a.s., for all large L,

$$|J_{(j)}^{(L)}| \geq 2|J_{(j+1)}^{(L)}| \quad \text{for } 1 \leq j < m_L. \tag{2.2}$$

This immediately implies that a.s., for all large L,

$$|J_{(j)}^{(L)}| > \sum_{k=j+1}^{m_L} |J_{(k)}^{(L)}| \quad \text{for } 1 \leq j < m_L. \tag{2.3}$$

17

The inequality (2.3) will play a crucial role in the structure of finite volume ground states.

There are many ways to arrange that (2.2) be valid for large L. Our way is to define

$$J_e^{(L)} = \alpha_e \exp(-\lambda^{(L)} K_e), \qquad (2.4)$$

where the K_e's are i.i.d. continuous random variables, the α_e's are i.i.d., independent of the K_e's, with $\nu(\alpha_e = +1) = \nu(\alpha_e = -1) = 1/2$, and the $\lambda^{(L)}$'s are positive constants. The K_e's are not restricted as to sign and in fact can have any continuous common distribution $\tilde{\mu}$, but we will sometimes find it convenient to take $\tilde{\mu}$ to be the uniform distribution on $(0, 1)$. The next proposition provides the choice of $\lambda^{(L)}$. We remark that it turns out that in some (but not all) respects the ground state structure of this highly disordered spin glass is equivalent to that of the corresponding highly disordered ferromagnet (where $\alpha_e \equiv +1$); the reader is referred to section 6 of [NS96a] for more details.

Proposition 2.1 *For any continuous $\tilde{\mu}$, there exists some choice of $\lambda^{(L)}$ such that a.s. (2.2) is valid for all large L.*

Proof. For any distinct e and e' in $\mathbb{E}^d$, the function

$$g(\lambda) = \nu\left(1/2 < e^{-\lambda K_e}/e^{-\lambda K_{e'}} < 2\right) = \nu\left(|K_e - K_{e'}| < \frac{\ln 2}{\lambda}\right) \qquad (2.5)$$

depends only on $\tilde{\mu}$ and tends to zero as $\lambda \to \infty$ by the continuity of $\tilde{\mu}$. For any particular L, the set of inequalities in (2.2) is not satisfied precisely when there are distinct e and e' touching Λ_L with $1/2 < |J_e^{(L)}|/|J_{e'}^{(L)}| < 2$. Thus,

$$\nu\left(|J_{(j)}^{(L)}| < 2|J_{(j+1)}^{(L)}| \text{ for some } j \in \{1, \ldots, m_{L-1}\}\right) \leq \binom{m_L}{2} g(\lambda^{(L)}). \qquad (2.6)$$

Since $g(\lambda) \to 0$ as $\lambda \to \infty$, we may choose $\lambda^{(L)}$ so large that

$$g(\lambda^{(L)}) \leq (m_L L)^{-2}. \qquad (2.7)$$

Then the LHS of (2.6) is bounded by L^{-2}, which is summable in L. The Borel-Cantelli Lemma then completes the proof.   □

Henceforth, we assume that $\tilde{\mu}$ is continuous and that $\lambda^{(L)}$ has been chosen, according to Proposition 2.1, so that a.s. (2.2) and (2.3) are valid for all large L. For such an L, we next give an algorithm for determining the finite volume ground states. The algorithm will construct for each x in Λ_L a tree graph $T^{(L)}(x)$ touching both x and a single site $w = w_L(x)$ in $\partial \Lambda_L$. These trees will depend on the ordering of the $|J_e^{(L)}|$'s and hence on the ordering of the K_e's, but not on the signs α_e nor on the b.c. $\bar{s}^{(L)}$. The ground state spin at x, $s_x^{(L)}$, will then be determined, as explained below, by $T^{(L)}(x)$ together with the signs α_e on the edges of $T^{(L)}(x)$ and the b.c. spin $\bar{s}_w^{(L)}$.

To obtain $T^{(L)}(x)$, we first construct a growing sequence of trees $T_0(x)$, $T_1(x), \ldots$, where $T_n(x)$ has $n+1$ sites $\{x_0, \ldots, x_n\}$ and n edges, as follows. $T_0(x)$ consists of the single site $x_0 = x$ and $T_{n+1}(x)$ is obtained from $T_n(x)$ by adjoining to it the site $x_{n+1} = y'$ and the edge $e = e_{n+1}(x) = \{x', y'\}$ with the largest value of $|J_e^{(L)}|$ (smallest value of K_e) among all the edges e in

$$\partial^* T_n(x) = \{\{x', y'\} \in \mathbb{E}^d : x' \in T_n(x), y' \notin T_n(x)\}. \tag{2.8}$$

Now, for $x \in \Lambda_L$, denote by $M_L(x)$ the smallest n such that $T_n(x)$ touches $\partial \Lambda_L$. Then $T^{(L)}(x)$ is defined to be $T_{M_L(x)}(x)$. For future use, note that the sequence $T_n(x)$ depends only on the relative order of the K_e's and does not depend on L. The next proposition gives the rest of the algorithm for determining finite volume ground states.

Proposition 2.2 *For a given L, suppose (2.3) is valid. For $x \in \Lambda_L$, define $r_L(x)$ to be the unique path in $T^{(L)}(x)$ connecting x and (the site $w_L(x)$ in) $\partial \Lambda_L$ and define*

$$\eta_L(x) = \prod_{e \in r_L(x)} sgn(J_e^{(L)}). \tag{2.9}$$

Then the ground state $s^{(L)}$ for $H_{\Lambda_L}^{\bar{s}^{(L)}}$ is given by

$$s_x^{(L)} = \eta_L(x)\bar{s}_{w_L(x)}^{(L)}, \text{ for } x \in \Lambda_L. \tag{2.10}$$

Proof. We claim that if the coupling magnitudes satisfy (2.3), then regardless of the coupling signs or the b.c., a ground state $s^{(L)}$ must have all bonds in all $T^{(L)}(x)$'s satisfied; i.e.,

$$\forall x \in \Lambda_L, \quad \forall \{x', y'\} \in T^{(L)}(x), \quad J_{\{x', y'\}}^{(L)} s_{x'}^{(L)} s_{y'}^{(L)} > 0. \tag{2.11}$$

Included in (2.11) is the case where $x' \in \Lambda_L$ and $y' \in \partial \Lambda_L$ (i.e., $y' = w_L(x)$), in which case $s_{y'}^{(L)}$ is replaced by the b.c. value $\bar{s}_{y'}^{(L)}$. The proposition is an immediate consequence of this claim.

To justify the claim, let $A_n(x)$ denote the set of sites in $T_n(x)$. For $n < M_L(x)$, $A_n(x) \subseteq \Lambda_L$ and $e_{n+1}(x)$ is the edge e in $\partial^* A_n(x) (= \partial^* T_n(x))$ with the largest value of $|J_e^{(L)}|$. If this edge were not satisfied for $s^{(L)}$, then flipping the spins in $A_n(x)$ would change the energy by an amount (see (1.1))

$$2 \sum_{\{x', y'\} \in \partial^* A_n(x)} J_{\{x', y'\}}^{(L)} s_{x'}^{(L)} s_{y'}^{(L)} = -2 \left(\left| J_{e_{n+1}(x)}^{(L)} \right| - \sum_{\substack{e = \{x', y'\} \in \partial^* A_n(x) \\ e \neq e_{n+1}(x)}} J_e^{(L)} s_{x'}^{(L)} s_{y'}^{(L)} \right)$$

$$\leq -2 \left(\left| J_{e_{n+1}(x)}^{(L)} \right| - {\sum}' \left| J_e^{(L)} \right| \right), \tag{2.12}$$

where $\sum'$ denotes the sum over those edges $e = \{x', y'\} \in \mathbb{E}^d$ with $e \neq e_{n+1}(x)$, with x' or y' in Λ_L and with $|J_e^{(L)}| \leq |J_{e_{n+1}(x)}^{(L)}|$. The RHS of (2.12) is strictly negative by (2.3), which would contradict $s^{(L)}$ being a ground state. This proves that for every x and every $n < M_L(x)$, $e_{n+1}(x)$ is satisfied in any ground state, which is just the claim made above. The proof is now complete. $\square$

Each tree $T^{(L)}(x)$ consists of sites and edges entirely in Λ_L except for the single site $w_L(x)$ in $\partial \Lambda_L$ and the edge $e_{M_L(x)}(x)$ connecting it to Λ_L. We now consider the graph

$$F^{(L)} = \bigcup_{x \in \Lambda_L} T^{(L)}(x), \tag{2.13}$$

or more precisely, we separately take the union of the vertices and of the edges of the $T^{(L)}(x)$'s. Obviously $F^{(L)}$ touches each $x \in \Lambda_L$, but not in general every $y \in \partial \Lambda_L$. Less obviously, it has no loops (simple closed paths) and so is a forest, i.e., a union of trees:

Proposition 2.3 *If the magnitudes $|J_e^{(L)}|$ for $e = \{x, y\}$ with x or y in Λ_L are distinct, then the graph $F^{(L)}$ is a union of disjoint trees (spanning all of Λ_L) with each tree touching $\partial \Lambda_L$ at a single site. Two sites x and x' in Λ_L belong to the same tree of $F^{(L)}$ if and only if (the vertex sets of) $T^{(L)}(x)$ and $T^{(L)}(x')$ intersect.*

Proof. Since the construction of the $T^{(L)}(x)$'s and $F^{(L)}$ depends only on the relative order of $|J_e^{(L)}|$'s and not on their actual values, we may assume (without loss of generality) that (2.3) is valid. If there were a loop in $F^{(L)}$, then there would be two distinct paths r_1 and r_2 in $F^{(L)}$ connecting two distinct points x and x' in Λ_L. Since $F^{(L)}$ does not depend on the signs of the coupling, we may assume (without loss of generality) that the product of the signs,

$$\eta_r(x, x') = \prod_{e \in r} \operatorname{sgn}(J_e^{(L)}), \tag{2.14}$$

is different for $r = r_1$ and $r = r_2$. But by (2.11), all the bonds in each path must be satisfied in any ground state $s^{(L)}$, which implies that $s_x^{(L)} \eta_r(x, x') s_{x'}^{(L)} = +1$ for $r = r_1$ and $r = r_2$. This contradiction shows that there are no loops. A similar argument shows that if there were a path r in $F^{(L)}$ connecting two distinct points w and w' in $\partial \Lambda_L$, then $\bar{s}_w^{(L)} \eta_r(w, w') \bar{s}_{w'}^{(L)} = +1$ for any ground state for any b.c. $\bar{s}$. Since $\eta_r(w, w')$ does not depend on the b.c., this would again lead to a contradiction, which shows that each tree in $F^{(L)}$ touches $\partial \Lambda_L$ at (one and) only one site. If $T^{(L)}(x)$ and $T^{(L)}(x')$ intersect, then clearly x and x' belong to the same tree; if $T^{(L)}(x)$ and $T^{(L)}(x')$ do not intersect then $w_L(x) \neq w_L(x')$ and so x and x' cannot belong to a tree which touches $\partial \Lambda_L$ at only one site. This completes the proof. $\square$

The ground state structure for finite L (large enough so that (2.3) is valid) is now clear: The forest $F^{(L)}$ is determined by the order of the coupling magnitudes, i.e., by the order of the K_e's. If x and x' belong to the same tree of $F^{(L)}$, then, regardless of the b.c., the relative sign $s_x^{(L)} s_{x'}^{(L)}$ in any ground state $s^{(L)}$, determined by $F^{(L)}$ and the coupling signs, $\text{sgn}(J_e^{(L)}) = \alpha_e$, is $\eta_r(x, x')$ of (2.14) with r the unique path connecting x and x' in the tree. The overall sign of each tree is determined (as in (2.10)) by the b.c. $\bar{s}^{(L)}$ according to its value at the unique site w in $\partial \Lambda_L$ touched by that tree. As $\bar{s}^{(L)}$ varies over all possible b.c.'s for the given L, for fixed K_e's and α_e's, one obtains all of the 2^{N_L} possible choices of overall signs for the N_L trees in $F^{(L)}$.

Since the sequence of growing trees $T_n(x)$ depends only on the (order of the) K_e's and not on L, there is, for each x in $\mathbb{Z}^d$, a well defined limiting infinite tree graph $T_\infty(x)$. We define

$$F_\infty = \bigcup_{x \in \mathbb{Z}^d} T_\infty(x), \qquad (2.15)$$

i.e., the graph with vertex set $\mathbb{Z}^d$ and every edge contained in any of the $T_\infty(x)$'s. This graph was also studied in [Ale95] as a $\mathbb{Z}^d$ version of a Poisson process based Euclidean model studied earlier in [AlS92]. Not suprisingly now, we have the following result from [NS94], [NS96a]; the first part, concerning F_∞ but not the spin glass model, also appears in [Ale95].

Theorem 2.4 *There is a number $\mathcal{N} = \mathcal{N}(d)$ in $\{1, \ldots, \infty\}$ such that, almost surely: F_∞ is a forest spanning all of $\mathbb{Z}^d$, consisting of $\mathcal{N}$ distinct trees, all infinite; x and x' belong to the same tree of F_∞ if and only if (the vertex sets of) $T_\infty(x)$ and $T_\infty(x')$ intersect; $s \in \mathcal{S}$ is an infinite volume ground state if and only if whenever x and x' are in the same tree of F_∞,*

$$s_x s_{x'} = \eta_\infty(x, x') = \prod_{e \in r_\infty(x, x')} \alpha_e, \qquad (2.16)$$

where $r_\infty(x, x')$ is the unique path in F_∞ connecting x and x'. Thus, the number of infinite volume ground states is a.s. $2^\mathcal{N}$ if $\mathcal{N} < \infty$ and a.s. uncountable if $\mathcal{N} = \infty$.

Proof. From the definition of $T^{(L)}(x)$, we have $T_\infty(x)$ as the increasing limit both of $T_n(x)$ and of $T^{(L)}(x)$ and F_∞ as the increasing limit of $F^{(L)}$. Since no $F^{(L)}$ has loops, neither does F_∞. F_∞ is clearly spanning and has all component trees infinite. $\mathcal{N}$ is invariant with respect to translations in $\mathbb{Z}^d$ (we assume here that $(\Omega, \mathcal{F}, \nu)$ is realized as the natural product space over $\mathbb{E}^d$) and so by ergodicity is a.s. a constant. Since $F^{(L)} \uparrow F_\infty$, x and x' are connected in F_∞ if and only if they are connected in some $F^{(L)}$. By part of Proposition 2.3, this is so if and only if $T^{(L)}(x)$ intersects $T^{(L)}(x')$ for some L. But this is equivalent to $T^\infty(x)$ intersecting $T^\infty(x')$. Thus, x and x' are in the same tree of F_∞ if and only if $T^\infty(x)$ intersects $T^\infty(x')$. The characterization of infinite volume ground states follows immediately from our characterization above of finite volume ground states and the fact that

$F^{(L)} \uparrow F_\infty$. Since the satisfaction of (2.16) within each tree of F_∞ leaves one overall choice of sign (e.g., for the spin value at one site) for that tree, the final claim of the theorem follows. $\square$

The stochastically growing vertex set of $T_n(x)$ is essentially the same as in (one of the standard versions of) a model called invasion percolation [LeB80], [CKLW82], [WiW83], that is closely related [WiW83], [ChCN85] to standard Bernoulli percolation on $\mathbb{Z}^d$, as we discuss below. One difference is that usual invasion percolation takes the distribution $\tilde{\mu}$ of each K_e to be uniform on $(0,1)$. Since all continuous $\tilde{\mu}$'s yield the same distribution for the relative order of the K_e's, this difference is entirely cosmetic. The second and main difference is that in the usual definition of invasion percolation, one also adds (or "invades") an edge $e = \{x', y'\}$ (with minimal K_e) even if x' and y' have already been reached, so that the resulting graph $\hat{T}_m(x)$ is not in general a tree. But if $\hat{M}_n$ denotes the first m such that $\hat{T}_m(x)$ contains n sites, then the vertex sets of $\hat{T}_{\hat{M}_n}(x)$ and $T_n(x)$ coincide, as do the vertex sets of $\hat{T}_\infty(x)$ and $T_\infty(x)$.

If, for a given dimension d, there were only a single pair of infinite volume ground states for the highly disordered spin glass of this chapter, that would be analogous to there being no nonconstant ground states for the disordered ferromagnet of the last chapter. By Theorem 2.4 and the discussion just above, the number of spin glass ground states is exactly two if and only if for all $x, x' \in \mathbb{Z}^d$,

$$\nu\left(\hat{T}_\infty(x) \cap \hat{T}_\infty(x') \neq \emptyset\right) = 1, \qquad (2.17)$$

i.e., $\hat{T}_\infty(x)$ and $\hat{T}_\infty(x')$ have at least one vertex in common. As in the analogous situation for the disordered ferromagnet, it is conjectured for this spin glass model [NS94] (as we explain later) that this is indeed so for low d. In both situations, the best results (for $d \neq 1$) are when $d = 2$. But in the spin glass case, the result is complete:

Theorem 2.5 ([ChCN85]) *For $d = 2$, the infinite invasion regions, $\hat{T}_\infty(x)$ and $\hat{T}_\infty(x')$, started from any two sites x and x' in $\mathbb{Z}^d$, a.s. intersect.*

Proof. The proof is based on the following natural coupling between invasion percolation on $\mathbb{Z}^d$ and standard (nearest neighbor) Bernoulli bond percolation on $\mathbb{Z}^d$, with open bond density p, and on special properties of the latter when $d = 2$ and $p = 1/2$. For the given $\tilde{\mu}$ and any $p \in (0,1)$, choose $\phi = \phi_p \in \mathbb{R}$ so that

$$\nu\left(K_e \leq \phi\right) = \tilde{\mu}\left((-\infty, \phi]\right) = p. \qquad (2.18)$$

Of course, if $\tilde{\mu}$ is uniform on $(0,1)$, then $\phi_p = p$. We then define an edge e to be open (for the chosen ϕ) if $K_e \leq \phi$, so that edges are open or closed as in the standard Bernoulli percolation with density p. The open cluster $C(x) (= C_\phi(x))$ of x is defined as the maximal connected graph (using vertices from $\mathbb{Z}^d$ and edges from $\mathbb{E}^d$) containing x and only using open edges. We denote by $|C(x)|$ the number of vertices in $C(x)$; it may be ∞.

It is clear that the invasion process $\{\hat{T}_m(x)\}$ only invades a closed edge when no open edge is available. Hence

$$|C(x)| < \infty \quad \Longrightarrow \quad \hat{T}_\infty(x) \supset C(x). \tag{2.19}$$

We also easily have, more generally, that

$$y \in \hat{T}_\infty(x) \text{ and } |C(y)| < \infty \quad \Longrightarrow \quad \hat{T}_\infty(x) \supset C(y). \tag{2.20}$$

It follows that $\hat{T}_\infty(x)$ and $\hat{T}_\infty(x')$ intersect if each touches some site in the same finite open cluster. This will a.s. be the case for $d = 2$, by taking $p = 1/2$ (which is the critical value for $d = 2$ [Kes80], although we will not use this) and using the classic result [Harr60] that for $d = 2$ and $p = 1/2$: a.s., for every L, there is an open cluster which is both finite and contains a loop enclosing Λ_L. This completes the proof. □

We remark that more detailed arguments in [ChCN85] show that Theorem 2.5 can be strengthened to yield that for $d = 2$ the symmetric difference of $\hat{T}_\infty(x)$ and $\hat{T}_\infty(x')$ is a.s. finite. Thus for $d = 2$, more than enough is known to conclude that the "invasion forest", F_∞, a.s. consists of a single tree. Since $\mathcal{N}(2) = 1$, according to Theorem 2.4, there are a.s. exactly two infinite volume ground states in our spin glass model for $d = 2$. In the absence of a rigorous determination of $\mathcal{N}(d)$ for $d > 2$, we discuss next some heuristic arguments [NS94], [NS96a], which suggest that $\mathcal{N}(d) = 1$ for $d < 8$ and $\mathcal{N}(d) = \infty$ for $d > 8$; the value of $\mathcal{N}(8)$ is unclear even on a heuristic level. We note that intuitively one expects $\mathcal{N}$ to be either 1 or ∞, but there seems to be no proof ruling out other values.

There are two types of heuristic reasoning, that together lead to the conjecture that $d = 8$ is the crossover dimension between $\mathcal{N} = 1$ and $\mathcal{N} = \infty$. Both concern the "dimension" D_i of $\hat{T}_\infty(x)$, say for $x = 0$. One possible "definition" of D_i is based on the idea that the number of sites of $\hat{T}_\infty(x)$ belonging to Λ_L is a.s. of order L^{D_i} (for some D_i), i.e., is bounded above and below by finite positive constants times L^{D_i}, as $L \to \infty$. A somewhat different version would be that L^{D_i} is the order of the mean number of sites, i.e., of the sum over y in Λ_L of $G_d(y)$, where

$$G_d(y - x) = \nu\left\{y \in \hat{T}_\infty(x)\right\}. \tag{2.21}$$

A stronger version would be that $G_d(y)$ itself is of order $||y||^{D_i - d}$.

The function G_d on $\mathbb{Z}^d$ is an important quantity in invasion percolation and plays a role analogous to that of the connectivity function in Bernoulli bond percolation, $\tau_p(x) = \tau_{p,d}(x)$, defined as

$$\tau_p(y - x) = \nu\left\{y \in C_p(x)\right\}. \tag{2.22}$$

To avoid overly nested subscripts, we take $\tilde{\mu}$ here and below to be uniform on $(0, 1)$ so that $\phi_p = p$. The conjecture that $G_d(x)$ is of order $||x||^{D_i - d}$ for some $D_i = D_i(d) < d$ is a natural one, which we will soon pursue a bit further, but it should be pointed out that $G_d(x)$ is not even proved to tend to zero as $||x|| \to \infty$ for every d (≥ 2). Indeed, it follows readily from (2.19) that

$$G_d(x) \geq \sup_{p<p_c(d)} \tau_{p,d}(x) = \tau_{p_c(d),d}(x), \qquad (2.23)$$

where $p_c = p_c(d)$ is the critical value of p (for having infinite clusters) on $\mathbb{Z}^d$. But by the uniqueness of infinite clusters for Bernoulli bond percolation [AKN87], [BuK89] and the FKG inequalities (for the case of independent Bernoulli variables [Harr60])

$$
\begin{aligned}
\tau_{p_c}(y-x) &\geq \nu\left(C_{p_c}(x) \text{ is infinite and the same as } C_{p_c}(y)\right) \\
&= \nu\left(|C_{p_c}(x)| = \infty, \ |C_{p_c}(y)| = \infty\right) \qquad (2.24) \\
&\geq \nu\left(|C_{p_c}(x)| = \infty\right) \nu\left(|C_{p_c}(y)| = \infty\right) = [\theta_d(p_c(d))]^2,
\end{aligned}
$$

where $\theta(p) = \theta_d(p)$ is the Bernoulli percolation order parameter,

$$\theta(p) = \nu\left(|C_p(x)| = \infty\right). \qquad (2.25)$$

Combining (2.23) and (2.24) shows that for any d,

$$\inf_{x \in \mathbb{Z}^d} G_d(x) = 0 \quad \Longrightarrow \quad \theta_d(p_c(d)) = 0. \qquad (2.26)$$

(In fact, the reverse implication can also be proved.) It is a well known open problem, that of "continuity of the percolation transition", to prove that $\theta_d(p_c(d)) = 0$ for general d. This has so far only been proved for $d = 2$ [Harr60], [Kes80] and for d large [HarS90], but by (2.26) this would be an implication of $G_d(x) \to 0$ as $||x|| \to \infty$!

Returning to the relation between $D_i(d)$ and $\mathcal{N}(d)$, the first piece of heuristic reasoning is that random D_i-dimensional objects (such as $\hat{T}_\infty(x)$ and $\hat{T}_\infty(x')$) in a (discrete) d-dimensional space ought to a.s. intersect if $D_i > d/2$ but should have a positive probability of nonintersecting if $D_i < d/2$. The second heuristic is somewhat more complicated and concerns the notion in the physics literature (see the references in [NS94], [NS96a]) that $D_i(d)$ equals the dimension $D^*(d)$ of another object (the so-called incipient infinite cluster) associated with critical Bernoulli percolation. This $D^*(d)$ is believed to grow with d (always being above $d/2$) for $d \leq 6$ and then stay fixed at the value 4 for $d \geq 6$. (See [A97] and [BoCKS97] for recent rigorous results related to this issue.) This of course leads to the crossover dimension value 8. We have nothing more to say here about this second piece of reasoning, but we now present a theorem which verifies half of the first piece of reasoning. In particular, it shows that if $G_d(x) = O\left(||x||^{D_i-d}\right)$ with $D_i < d/2$, then $\hat{T}_\infty(x)$ and $\hat{T}_\infty(x')$ need not intersect and in fact $\mathcal{N}(d) = \infty$.

Theorem 2.6 ([NS94], [NS96a]) *For any d,*

$$\sum_{x \in \mathbb{Z}^d} [G_d(x)]^2 < \infty \quad \Longrightarrow \quad \mathcal{N}(d) = \infty. \qquad (2.27)$$

Proof. We will work with $\{\hat{T}_m(x)\}$ rather than with $\{T_n(x)\}$; the arguments are the same for the two cases. The main point is to prove that square summability of G_d implies that

$$\nu\left(\hat{T}_\infty(x) \cap \hat{T}_\infty(y) = \emptyset\right) \to 1 \quad \text{as} \quad ||x - y|| \to \infty. \tag{2.28}$$

Once we have this, then for any large n, we can find $x_1, \ldots, x_n \in \mathbb{Z}^d$ with $||x_i - x_j||$ sufficiently large for each $i < j$ so that

$$\nu\left(\hat{T}_\infty(x_i) \cap \hat{T}_\infty(x_j) = \emptyset \text{ for each } i < j\right)$$
$$\geq 1 - \sum_{i<j} \nu\left(\hat{T}_\infty(x_i) \cap \hat{T}_\infty(x_j) \neq \emptyset\right) > 0. \tag{2.29}$$

Then, from Theorem 2.4, we have that $\mathcal{N}(d) \geq n$ for every n, so $\mathcal{N}(d) = \infty$.

If $\hat{T}_\infty(x)$ and $\hat{T}_\infty(y)$ were independent random objects, then to prove (2.28), we could mimic the proof that two independent random walk paths in $\mathbb{Z}^d$ starting from well-separated x and y do not intersect for $d > 4$, with probability close to 1, because of the square summability of $\tilde{G}_d(x)$, the probability that a random walk in $\mathbb{Z}^d$ starting at the origin ever visits x. The main idea of our proof is that this argument still works despite the dependence of $\hat{T}_\infty(x)$ and $\hat{T}_\infty(y)$ because, roughly speaking, they are independent as long as they don't almost intersect.

More precisely, let $\left\{\hat{T}'_m(x) : m \geq 0, x \in \mathbb{Z}^d\right\}$ and $\left\{\hat{T}''_m(x) : m \geq 0, x \in \mathbb{Z}^d\right\}$ be defined on some probability space $(\Omega^*, \mathcal{F}^*, P^*)$ such that each is equidistributed with $\{\hat{T}_m(x)\}$ and the two are independent. For two graphs A and B on $(\mathbb{Z}^d, \mathbb{E}^d)$ let $\rho(A, B)$ denote the minimum Euclidean distance between any vertex of A and any vertex of B. We claim that

$$\nu\left(\rho\left(\hat{T}_\infty(x), \hat{T}_\infty(y)\right) > 1\right) = P^*\left(\rho\left(\hat{T}'_\infty(x), \hat{T}''_\infty(y)\right) > 1\right). \tag{2.30}$$

Once we have justified this claim, we will then indeed mimic the random walk argument to prove that the RHS of (2.30) tends to 1 as $||x - y|| \to \infty$, which will imply (2.28), as desired.

Here is a concrete argument for (2.30). For each $m = 0, 1, 2, \ldots$, we consider the finitely many possible outcomes A for $\hat{T}_m(x)$ and $\hat{T}'_m(x)$ and B for $\hat{T}_m(y)$ and $\hat{T}''_m(y)$. It should be clear from the definition of invasion percolation that an event like $\{\hat{T}_m(x) = A\}$ belongs to the σ-field generated by

$$\{K_e : e \text{ touches at least one vertex in } A\}, \tag{2.31}$$

and similarly for $\{\hat{T}_m(y) = B\}$. Hence

$$\rho(A, B) > 1 \quad \Longrightarrow \quad \text{independence of } \{\hat{T}_m(x) = A\} \text{ and } \{\hat{T}_m(y) = B\}. \tag{2.32}$$

Thus letting $\sum^1$ denote the sum over A, B with $\rho(A, B) > 1$, we have

$$
\begin{aligned}
\nu\left(\rho(\hat{T}_m(x), \hat{T}_m(y)) > 1\right) &= \sum\nolimits^1 \nu\left(\hat{T}_m(x) = A,\ \hat{T}_m(y) = B\right) \\
&= \sum\nolimits^1 P^*\left(\hat{T}'_m(x) = A,\ \hat{T}''_m(y) = B\right) \quad (2.33) \\
&= P^*\left(\rho(\hat{T}'_m(x), \hat{T}''_m(y)) > 1\right).
\end{aligned}
$$

Letting $m \to \infty$ yields (2.30).

We proceed to obtain a lower bound for the RHS of (2.30). For $z \in \mathbb{Z}^d$, let $\hat{I}_z$ denote the event that $z \in \hat{T}'_\infty(x)$ and $\rho(z, \hat{T}''_\infty(y)) \leq 1$. Then

$$
\begin{aligned}
P^*\left(\rho(\hat{T}'_\infty(x), \hat{T}''_\infty(y)) > 1\right) &= 1 - P^*\left(\bigcup_{z \in \mathbb{Z}^d} \hat{I}_z\right) \\
&\geq 1 - \sum_{z \in \mathbb{Z}^d} P^*\left(\hat{I}_z\right) \quad (2.34) \\
&= 1 - \sum_{z \in \mathbb{Z}^d} P^*\left(z \in \hat{T}'_\infty(x)\right) P^*\left(\rho(z, \hat{T}''_\infty(y)) \leq 1\right) \\
&\geq 1 - \sum_{z \in \mathbb{Z}^d} G_d(z - x) \sum_{z': \|z' - z\| \leq 1} G_d(z' - y).
\end{aligned}
$$

If we let $\tilde{z} = x - z$ and $w = z' - z$, and use the fact that $G_d(-\tilde{z}) = G_d(\tilde{z})$, then the last expression equals

$$
= 1 - \sum_{w: \|w\| \leq 1} \sum_{\tilde{z} \in \mathbb{Z}^d} G_d(x - y + w - \tilde{z}) G_d(\tilde{z}) = 1 - \sum_{w: \|w\| \leq 1} (G_d * G_d)(x - y + w),
$$
(2.35)

where the convolution $G * G$ is

$$
(G * G)(x) = \sum_{z \in \mathbb{Z}^d} G(x - z) G(z). \quad (2.36)
$$

To complete the proof, it suffices to show $(G * G)(x) \to 0$ as $\|x\| \to \infty$. But the square summability of G implies that its Fourier transform

$$
\hat{G}(k) = (2\pi)^{-d/2} \sum_{x \in \mathbb{Z}^d} G(x) e^{i(k,x)} \quad (2.37)
$$

is in $L^2\left([-\pi, \pi]^d, dk\right)$, where (k, x) denotes the standard inner product and dk denotes Lebesgue measure. Furthermore,

$$
(G * G)(x) = \int_{[-\pi, \pi]^d} \left[\hat{G}(k)\right]^2 e^{-i(k,x)} dk, \quad (2.38)
$$

and so $(G * G)(x) \to 0$ as $\|x\| \to \infty$ by the Riemann-Lebesgue lemma. The proof is now complete. $\qquad\square$

Chapter 3

High Temperature States of Disordered Systems

We return now to the more realistic models where $J_e^\Lambda = J_e$, with no dependence on Λ. The couplings J_e, for $e \in \mathbb{E}^d$, are i.i.d. random variables on $(\Omega, \mathcal{F}, \nu)$ with common distribution μ. Since we will not be considering ground states here, we do not require μ to be continuous. Indeed, except where otherwise noted, we make no particular restriction on μ. Ferromagnetic (where μ is supported on $[0, \infty)$) and spin glass (where μ is symmetric) models are of course important special cases. The topic of this chapter is a percolation-based approach to obtaining bounds for β_c, the critical value of β (or equivalently bounds for the critical temperature), defined here as the supremum of β's for which there is a unique infinite volume Gibbs distribution (regardless of b.c.'s) for all inverse temperatures $\leq \beta$.

We will present two types of results. One type, from [ACCN87], concerns ferromagnetic models and gives both upper and lower bounds, by relating (via inequalities) disordered ferromagnets to standard homogeneous Bernoulli bond percolation. The other type of result, from [N94], uses percolation methods to show that for a given β, an Ising model with couplings J_e (not necessarily random) of varying sign has a unique infinite volume Gibbs distribution if that is so for the associated ferromagnet with couplings $|J_e|$. Applied to a disordered model, this bounds β_c below (for the model with common distribution μ) by the critical value β_c^F for the associated disordered ferromagnet with μ replaced by μ^F, the common distribution of the $|J_e|$'s. The simplest special case is the $\pm \check{J}$ spin glass, where $J_e = +\check{J}$ or $-\check{J}$ with probability $1/2$; here the associated ferromagnet is just the standard homogeneous Ising model. We remark that it seems to be an open problem even to slightly improve $\beta_c \geq \beta_c^F$ to a strict inequality, $\beta_c > \beta_c^F$. We also note that some of the results presented here are included in a recent survey [H96] and most of the results are rederived in [AleC96].

Before introducing our percolation-based approach, we first briefly review some facts about (finite and) infinite volume Gibbs states. For a finite $\Lambda \subset \mathbb{Z}^d$, a b.c., more general than fixing $\bar{s} \in \mathcal{S}_{\partial \Lambda} = \{-1, +1\}^{\partial \Lambda}$, may be specified by a probability measure $\bar{\rho}$ on $\mathcal{S}_{\partial \Lambda}$. The corresponding Gibbs distribution on $\mathcal{S}_\Lambda$, denoted by $P_{\Lambda,\beta}^{\bar{\rho}}$, is the mixture of fixed b.c. Gibbs distributions,

$$P_{\Lambda,\beta}^{\bar{\rho}} = \sum_{\bar{s} \in \mathcal{S}_{\partial\Lambda}} \bar{\rho}(\{\bar{s}\}) P_{\Lambda,\beta}^{\bar{s}}. \tag{3.1}$$

Later on, we will need to consider other types of b.c.'s. Meanwhile, we define an infinite volume Gibbs state (at inverse temperature β) for given $\{J_e : e \in \mathbb{E}^d\}$ as any measure P_β on $\mathcal{S} = \{-1, +1\}^{\mathbb{Z}^d}$ such that there is some sequence of b.c.'s (on the cubes Λ_L), i.e., $\bar{\rho}_L$ on $\mathcal{S}_{\partial\Lambda_L}$, such that P_β is the limit (in the sense of convergence of finite dimensional distributions) of $P_{\Lambda_L,\beta}^{\bar{\rho}_L}$ as $L \to \infty$. We remark that there is no loss of generality resulting from the restriction to cubes or to this class of b.c.'s.

The intrinsic characterization of P_β is that it satisfies the Dobrushin-Lanford-Ruelle (DLR) equations (see, e.g., [Ge88] for a more complete discussion and for historical references), i.e., for every finite Λ, the conditional distribution of P_β, conditioned on the σ-field generated by $\{s_x : x \in \mathbb{Z}^d \setminus \Lambda\}$, is $P_{\Lambda,\beta}^{\bar{s}}$ where $\bar{s} \in \mathcal{S}_{\partial\Lambda}$ is given by the (conditioned) values of s_x for $x \in \partial\Lambda$. We will make occasional use of this characterization and also of the following elementary fact, whose proof we leave to the reader.

Proposition 3.1 *For given J_e's and β, there is a unique infinite volume Gibbs state if and only if for any two sequences of fixed boundary conditions $\bar{s}^{(L)}$ and $\bar{t}^{(L)}$ on $\partial\Lambda_L$, the limits as $L \to \infty$ of $P_{\Lambda_L,\beta}^{\bar{s}^{(L)}}$ and $P_{\Lambda_L,\beta}^{\bar{t}^{(L)}}$ exist and are the same P_β.*

Our approach to uniqueness and nonuniqueness of Gibbs states for Ising models is based on the Fortuin-Kasteleyn (FK) random cluster model representation. Although this was originally constructed only for the case of ferromagnets [KasF69], [ForK72] (see also [ACCN88], [Gri94]), it has a straightforward extension to Ising models with arbitrary coupling signs [KaO88] (see also [SW87], [ES88], [N91], [N94]).

For Λ a (finite) subset of $\mathbb{Z}^d$, we denote by $\hat{\Lambda}$ the set $e = \{x, y\}$ in $\mathbb{E}^d$ with both x and y in Λ. The Gibbs distribution $P_{\Lambda,\beta}^f$ is a probability measure on $\mathcal{S}_\Lambda = \{-1, +1\}^\Lambda$ and the FK model distribution will be a probability measure $\hat{P}_{\hat{\Lambda},\beta}^f$ on $\mathcal{N}_{\hat{\Lambda}} = \{0, 1\}^{\hat{\Lambda}}$. We regard $n \in \mathcal{N}_{\hat{\Lambda}}$ as specifying a percolation configuration of open ($n_e = 1$) and closed ($n_e = 0$) edges in $\hat{\Lambda}$. We will call the open clusters of that configuration n-clusters. A particularly aesthetic way of constructing $\hat{P}_{\hat{\Lambda},\beta}^f$ and elucidating its relation to $P_{\Lambda,\beta}^f$ is to obtain it as the marginal distribution (on $\mathcal{N}_{\hat{\Lambda}}$) of a joint distribution, $\tilde{P}_\beta^f = \tilde{P}_{\Lambda,\hat{\Lambda},\beta}^f$, on $\mathcal{S}_\Lambda \times \mathcal{N}_{\hat{\Lambda}}$, constructed in two steps, as follows.

Step 1. Let $P_\beta' = P_{\Lambda,\hat{\Lambda},\beta}'$ be the distribution on $\mathcal{S}_\Lambda \times \mathcal{N}_{\hat{\Lambda}}$ of random variables $\{S_x', N_e' : x \in \Lambda, e \in \hat{\Lambda}\}$ which are mutually independent and such that $S_x' = +1$ or -1 with probability $1/2$ while $N_e' = 1$ (resp., 0) with probability p_e (resp., $1 - p_e$), where

$$p_e = 1 - e^{-2\beta|J_e|}. \tag{3.2}$$

Step 2. Define

$$\tilde{U} = \tilde{U}^f_{\Lambda,\hat{\Lambda}} = \left\{(s,n) \in \mathcal{S}_\Lambda \times \mathcal{N}_{\hat{\Lambda}} : \forall e = \{x,y\} \in \hat{\Lambda},\ n_e J_e s_x s_y \geq 0\right\}; \qquad (3.3)$$

this is the event that every open edge e (with $J_e \neq 0$) is satisfied. Let $\tilde{P}^f_\beta$ be P'_β conditioned on $\tilde{U}$; i.e.,

$$\tilde{P}^f_\beta (\{(s,n)\}) = \left[P'_\beta(\tilde{U})\right]^{-1} 1_{\tilde{U}}((s,n)) P'_\beta (\{(s,n)\}). \qquad (3.4)$$

Proposition 3.2 *The marginal distribution on $\mathcal{S}_\Lambda$ of $\tilde{P}^f_\beta$ is the Gibbs distribution $P^f_{\Lambda,\beta}$. The marginal distribution on $\mathcal{N}_{\hat{\Lambda}}$ of $\tilde{P}^f_\beta$, denoted by $\hat{P}^f_{\Lambda,\beta}$, has the form*

$$\hat{P}^f_{\Lambda,\beta}(\{n\}) = \left(\hat{Z}^f_{\Lambda,\beta}\right)^{-1} 1_U(n)\, 2^{\#^f(n)}\, \hat{P}^{ind}_{\Lambda,\beta}(\{n\}), \qquad (3.5)$$

where $\hat{P}^{ind}_{\Lambda,\beta}$ is the product measure on $\mathcal{N}_{\hat{\Lambda}}$ corresponding to independent n_e's with $\hat{P}^{ind}_{\Lambda,\beta}(n_e = 1) = p_e$ for each $e \in \hat{\Lambda}$,

$$\#^f(n) = \quad \text{the number of } n\text{-clusters}, \qquad (3.6)$$
$$U = \left\{n \in \mathcal{N}_{\hat{\Lambda}} : \exists \text{ some } s \in \mathcal{S}_\Lambda \text{ such that } (s,n) \in \tilde{U}\right\}, \qquad (3.7)$$

and $\hat{Z}^f_{\Lambda,\beta}$ is a normalization constant (so that $\hat{P}^f_{\Lambda,\beta}$ is a probability measure). The conditional distribution on $\mathcal{S}_\Lambda$ of $\tilde{P}^f_\beta$, conditioned on the percolation configuration $n \in \mathcal{N}_{\hat{\Lambda}}$, corresponds to choosing one site x from every n-cluster, taking those s_x's as i.i.d. with probability $1/2$ for the value $+1$ or -1, and then (uniquely) determining every other $s_{x'}$ so that $(s,n) \in \tilde{U}$. The conditional distribution on $\mathcal{N}_{\hat{\Lambda}}$ of $\tilde{P}^f_\beta$, conditioned on the spin configuration $s \in \mathcal{S}_\Lambda$, corresponds to taking the n_e's for satisfied edges e (i.e., where $e = \{x,y\}$ with $J_e s_x s_y \geq 0$) as independent random variables with probability p_e (resp., $1-p_e$) for the value 1 (resp., 0), and setting $n_{e'} = 0$ for every unsatisfied edge (so that $(s,n) \in \tilde{U}$).

Proof. All the claims of the proposition follow from elementary calculations. We only mention a few key points. In determining the marginal on $\mathcal{S}_\Lambda$, one sums (3.4) over all n for s fixed. Because of $1_{\tilde{U}}$ that sum is really over n such that $(s,n) \in \tilde{U}$. For each satisfied e, both $n_e = 0$ and $n_e = 1$ are included in the sum, resulting in a factor $(1-p_e) + p_e = 1$, but for an unsatisfied e, only $n_e = 0$ is allowed, resulting in a factor $1 - p_e = e^{-2\beta|J_e|}$. This yields $P^f_{\Lambda,\beta}$ because

$$\sum_{\text{unsatisfied } e \in \hat{\Lambda}} 2|J_e| = \sum_{e=\{x,y\} \in \hat{\Lambda}} |J_e|(-\text{sgn}(J_e)s_x s_y + 1) = H^f_\Lambda(s) + \text{const.} \qquad (3.8)$$

In determining the marginal on $\mathcal{N}_{\hat{\Lambda}}$, the sum is really over s such that $(s,n) \in \tilde{U}$. If $n \notin U$, there is no such s; this yields the 1_U in (3.5). If $n \in U$, then in each n-cluster, the definition of $\tilde{U}$ fixes the relative sign of all spin variables but leaves two choices of overall sign; this yields the $2^{\#^f(n)}$ in (3.5). This completes our discussion of the proof. $\square$

We remark that for a ferromagnetic model, the structure given in Proposition 3.2 simplifies considerably. For any n, the choice of $s \equiv +1$ or $s \equiv -1$ provides an $(s,n) \in \tilde{U}$; thus $U = \mathcal{N}_{\hat{\Lambda}}$ and 1_U can be deleted from (3.5). Similarly, the demand that $(s,n) \in \tilde{U}$ requires simply that in each n-cluster, the spins have a common sign. This gives a particularly simple form for the conditional on $\mathcal{S}_\Lambda$, given n, and thus for the relation between $P_{\Lambda,\beta}^{f,F}$ and $\hat{P}_{\hat{\Lambda},\beta}^{f,F}$ (where the superscript F is for ferromagnetic), which implies, for example, that for $x,y \in \Lambda$,

$$E_{\Lambda,\beta}^{f,F}(S_x S_y) = \hat{P}_{\hat{\Lambda},\beta}^{f,F}(x \xleftrightarrow{n} y), \tag{3.9}$$

where the LHS expectation is w.r.t. $P_{\Lambda,\beta}^{f,F}$ and $x \xleftrightarrow{n} y$ denotes the event that x and y are in the same n-cluster. The analogous relation for general signs of couplings is the somewhat more complicated

$$E_{\Lambda,\beta}^{f}(S_x S_y) = \hat{E}_{\hat{\Lambda},\beta}^{f}(\eta_n(x,y)), \tag{3.10}$$

where $\eta_n(x,y)$ (compare with (2.14)) is defined as

$$\eta_n(x,y) = \begin{cases} \prod_{e \in r} \operatorname{sgn}(J_e) & , \quad \text{if } x \xleftrightarrow{n} y, \\ 0 & , \quad \text{otherwise,} \end{cases} \tag{3.11}$$

with r any open n-path connecting x and y. The definition of U ensures that $\eta_n(x,y)$ does not depend on the choice of r.

For a fixed b.c. $\bar{s}$ on $\partial\Lambda$, we replace $\hat{\Lambda}$ by

$$\hat{\Lambda}^* = \left\{ \{x,y\} \in \mathbb{E}^d : x \text{ or } y \text{ or both are in } \Lambda \right\}, \tag{3.12}$$

and define an FK measure $\hat{P}_{\hat{\Lambda}^*,\beta}^{\bar{s}}$ on $\mathcal{N}_{\hat{\Lambda}^*}$. The construction is analogous to the free b.c. case, but with some differences. It is convenient to begin in the first step with the product measure $P_{\Lambda^*,\hat{\Lambda}^*,\beta}'$ on $\mathcal{S}_{\Lambda^*} \times \mathcal{N}_{\hat{\Lambda}^*}$ where $\Lambda^* = \Lambda \cup \partial\Lambda$. We then define $\tilde{P}_\beta^{\bar{s}} = \tilde{P}_{\Lambda^*,\hat{\Lambda}^*,\beta}^{\bar{s}}$ by conditioning on

$$\tilde{U}^{\bar{s}} = \tilde{U}_{\Lambda^*,\hat{\Lambda}^*}^{\bar{s}} = \left\{ (s,n) \in \mathcal{S}_{\Lambda^*} \times \mathcal{N}_{\hat{\Lambda}^*} : \forall x \in \partial\Lambda,\ s_x = \bar{s}_x \text{ and} \right.$$

$$\left. \forall e = \{x,y\} \in \hat{\Lambda}^*,\ n_e J_e s_x s_y \geq 0 \right\}. \tag{3.13}$$

The analogue of Proposition 3.2 for fixed b.c.'s is as follows; we leave the details of the analogous proof to the reader.

Proposition 3.3 *The marginal on $\mathcal{S}_{\Lambda^*}$ of $\tilde{P}_\beta^{\bar{s}}$ is the product of the point measure $\delta_{\bar{s}}$ on $\mathcal{S}_{\partial\Lambda}$ with the Gibbs measure $P_{\Lambda,\beta}^{\bar{s}}$ on $\mathcal{S}_\Lambda$. The marginal on $\mathcal{N}_{\hat{\Lambda}^*}$ is*

$$\hat{P}_{\hat{\Lambda}^*,\beta}^{\bar{s}}(\{n\}) = \left(\hat{Z}_{\hat{\Lambda}^*,\beta}^{\bar{s}}\right)^{-1} 1_{U^{\bar{s}}}(n) \, 2^{\#^{\bar{s}}(n)} \, \hat{P}_{\hat{\Lambda}^*,\beta}^{ind}(\{n\}), \tag{3.14}$$

where $\hat{P}_{\hat{\Lambda}^*,\beta}^{ind}(\{n\})$ is the product measure with the parameters p_e for $e \in \hat{\Lambda}^*$,

$$\#^{\bar{s}}(n) = \quad \text{the number of } n\text{-clusters not touching } \partial\Lambda, \tag{3.15}$$

$$U^{\bar{s}} = U_{\hat{\Lambda}^*}^{\bar{s}} = \left\{n \in \mathcal{N}_{\hat{\Lambda}^*} : \exists \text{ some } s \in \mathcal{S}_{\Lambda^*} \text{ such that } (s,n) \in \tilde{U}^{\bar{s}}\right\}, \tag{3.16}$$

and $\hat{Z}_{\hat{\Lambda}^*,\beta}^{\bar{s}}$ is a normalization constant. The conditional on $\mathcal{S}_{\Lambda^*}$, given $n \in \mathcal{N}_{\hat{\Lambda}^*}$, corresponds to setting $s_x = \bar{s}_x$ for every $x \in \partial\Lambda$, then setting s_x, for every x in an n-cluster touching $\partial\Lambda$, to the unique value required for (s,n) to be in $\tilde{U}^{\bar{s}}$, then choosing the relative signs for spins in each n-cluster not touching $\partial\Lambda$ to the values required by $\tilde{U}^{\bar{s}}$ and finally choosing the overall signs for these latter clusters by independent flips of a fair coin. The conditional on $\mathcal{N}_{\hat{\Lambda}^*}$, given $s \in \mathcal{S}_{\hat{\Lambda}^*}$, is as in Proposition 3.2.

The sets $\tilde{U}$ and $\tilde{U}^{\bar{s}}$ of the last two propositions correspond to the events (in the product space of joint spin and percolation configurations) that all open edges are satisfied. The sets U and $U^{\bar{s}}$ (in the space of percolation configurations alone) correspond to the events that the open edges are "unfrustrated", i.e., can be all simultaneously satisfied. We remark that even for a ferromagnetic model, for most choices of $\bar{s}$, $U^{\bar{s}}$ is *not* all of $\mathcal{N}_{\hat{\Lambda}^*}$ since open n-paths connecting boundary sites with different values of $\bar{s}_x$ are forbidden. Of course, in the special case of $\bar{s} \equiv +1$ or $\bar{s} \equiv -1$, $U^{\bar{s}}$ is all of $\mathcal{N}_{\hat{\Lambda}^*}$. These two b.c.'s give the same FK measure (in general, $\bar{s}$ and $-\bar{s}$ give the same FK measure) which in the ferromagnetic case is called the "wired" b.c. FK measure and denoted $\hat{P}_{\hat{\Lambda}^*,\beta}^w$. We also note that for fixed b.c.'s, the function $\#^{\bar{s}}(n)$ on $\mathcal{N}_{\hat{\Lambda}^*}$ is the same for all $\bar{s}$ (and we write $\#^w(n)$); this will not be so for some other b.c.'s we will need to consider later. Finally, we note that replacing $\#^{\bar{s}}(n) = \#^w(n)$ in (3.14) by $\#^w(n) + 1$ (and adjusting the normalization constant $\hat{Z}$ accordingly) leaves the measure unchanged; $\#^w(n) + 1$ is the number of n-clusters, where the clusters touching $\partial\Lambda$ are together counted as a single cluster.

For a ferromagnet with $\bar{s} \equiv +1$ or $\bar{s} \equiv -1$, we again have a very simple relation between the Gibbs measures $P_{\Lambda,\beta}^{\pm,F}$ and the FK measure $\hat{P}_{\hat{\Lambda}^*,\beta}^{w,F}$ which implies, in particular, that for $x \in \Lambda$,

$$E_{\Lambda,\beta}^{\pm,F}(S_x) = \pm\hat{P}_{\hat{\Lambda}^*,\beta}^{w,F}(x \xleftrightarrow{n} \partial\Lambda), \tag{3.17}$$

where $x \xleftrightarrow{n} \partial\Lambda$ denotes the event that the n-cluster of x touches $\partial\Lambda$. For general signs of couplings and/or a general fixed b.c. $\bar{s}$, the analogue is

$$E_{\Lambda,\beta}^{\bar{s}}(S_x) = \hat{E}_{\hat{\Lambda}^*,\beta}^{\bar{s}}(\eta_n(x)), \tag{3.18}$$

where $\eta_n(x)$ (compare with (2.10)) is defined as

$$\eta_n(x) = \begin{cases} \left[\prod_{e \in r} \operatorname{sgn}(J_e)\right] \bar{s}_z & , \quad \text{if } x \xleftrightarrow{n} \partial\Lambda, \\ 0 & , \quad \text{otherwise}, \end{cases} \qquad (3.19)$$

with r any open n-path connecting x and some site z in $\partial\Lambda$. The definition of $U^{\bar{s}}$ ensures that $\eta_n(x)$ does not depend on the choice of either z or r.

For the particular applications of the FK representation given below, we need some comparison inequalities. These will be based on FKG inequalities, as we now discuss. For certain probability measures $\hat{P}$ on $\mathcal{N}_{\hat{\Lambda}^*}$, it is the case that coordinate-wise nondecreasing real functions $f(n)$ and $g(n)$ (we say that f, g are increasing or write $f, g \uparrow$) are positively correlated, i.e.,

$$f, g \text{ increasing} \quad \Longrightarrow \quad \hat{E}(fg) \geq \hat{E}(f)\hat{E}(g). \qquad (3.20)$$

This was first shown in [Harr60] for a product measure. It was later extended [S69], [ForKG71] to more general measures, satisfying

$$\forall n, n' \in \mathcal{N}_{\hat{\Lambda}^*}, \quad \hat{P}(\{n \wedge n'\})\hat{P}(\{n \vee n'\}) \geq \hat{P}(\{n\})\hat{P}(\{n'\}), \qquad (3.21)$$

where $\{n \wedge n'\}_e = \min(n_e, n'_e)$ and $\{n \vee n'\}_e = \max(n_e, n'_e)$. The condition (3.21) (which is satisfied as an equality for a product measure) is often called the FKG lattice condition while the inequalities of (3.20) are called the FKG inequalities. The following fact is both trivial and extremely useful.

Proposition 3.4 *If (3.20) is valid for $\hat{P}$, and $\hat{P}'$ is defined by*

$$\hat{P}'(\{n\}) = g(n)\hat{P}(\{n\})/\hat{E}(g), \qquad (3.22)$$

where g is non-negative (with $\hat{E}(g) > 0$) and $\uparrow$ (resp., $\downarrow$), then $\hat{P} \ll \hat{P}'$ (resp., $\hat{P} \gg \hat{P}'$), i.e.,

$$f \text{ increasing} \quad \Longrightarrow \quad \hat{E}(f) \leq \hat{E}'(f) \quad (\text{resp., } \hat{E}(f) \geq \hat{E}'(f)). \qquad (3.23)$$

Proof. Since $\hat{E}'(f) = \hat{E}(fg)/\hat{E}(g)$, the $g \uparrow$ case follows immediately from (3.20). The $g \downarrow$ case is also immediate by letting $h = -g$ (which is $\uparrow$) and then applying (3.20) with g replaced by h. $\qquad \square$

As a corollary of this proposition and the original FKG inequalities of Harris [Harr60], we have the following proposition [For72] (see also [ACCN87]) about FK measures coming from ferromagnetic couplings, where we replace the subscript β by the subscript (p_e) to emphasize that the measure is determined by the p_e's (see (3.2)).

Proposition 3.5 *For the ferromagnetic FK measures with wired and free b.c.'s, given by*

$$\hat{P}^{w,F}_{\hat{\Lambda}^*,(p_e)}(\{n\}) = \left(\hat{Z}^{w,F}_{\hat{\Lambda}^*,(p_e)}\right)^{-1} 2^{\#^w(n)} \hat{P}^{ind}_{\hat{\Lambda}^*,(p_e)}(\{n\}),$$

$$\hat{P}^{f,F}_{\hat{\Lambda},(p_e)}(\{n\}) = \left(\hat{Z}^{f,F}_{\hat{\Lambda},(p_e)}\right)^{-1} 2^{\#^f(n)} \hat{P}^{ind}_{\hat{\Lambda},(p_e)}(\{n\}),$$

$$\qquad (3.24)$$

we have the comparison inequalities,

$$\hat{P}^{ind}_{\hat{\Lambda}^*,(h(p_e))} \quad \ll \quad \hat{P}^{w,F}_{\hat{\Lambda}^*,(p_e)} \quad \ll \quad \hat{P}^{ind}_{\hat{\Lambda}^*,(p_e)},$$

$$\hat{P}^{ind}_{\hat{\Lambda},(h(p_e))} \quad \ll \quad \hat{P}^{f,F}_{\hat{\Lambda},(p_e)} \quad \ll \quad \hat{P}^{ind}_{\hat{\Lambda},(p_e)},$$

(3.25)

where

$$h(p_e) = \frac{p_e}{2(1 - p_e) + p_e}.$$

(3.26)

Proof. We apply the original FKG inequalities of Harris and Proposition 3.4. We give the arguments for wired b.c.'s; they are the same for free b.c.'s. The right-hand inequality follows because $\#^w(n)$ (and hence $2^{\#^w(n)}$) is $\downarrow$. The left-hand inequality follows because

$$dP^{w,F}_{\hat{\Lambda}^*,(p_e)} \; / \; d\hat{P}^{ind}_{\hat{\Lambda}^*,(h(p_e))} \quad = \quad \text{const. } 2^{\#^w(n)} \prod_{e \in \hat{\Lambda}^*} \left[\frac{1 - h(p_e)}{h(p_e)} \cdot \frac{p_e}{1 - p_e} \right]^{n_e}$$

$$= \quad \text{const. } 2^{\#^w(n)} \prod_{e \in \hat{\Lambda}^*} 2^{n_e}$$

(3.27)

is increasing. This is so because if any $n_{e'}$ is increased from 0 to 1, $\#^w(n)$ either stays the same or decreases by 1 while $\sum_{e \in \hat{\Lambda}^*} n_e$ definitely increases by 1. $\quad\square$

For nonferromagnetic FK measures, we only have a one-sided inequality, given in the next proposition (see (3.29)). For the nonferromagnetic case, we use the subscript (J_e) to emphasize that the measure depends both on (p_e) (i.e., on the magnitudes $\beta|J_e|$) and on the signs of the J_e's.

Proposition 3.6 *For the wired and free ferromagnetic FK measures, $\hat{P}^{w,F}_{\hat{\Lambda}^*,(p_e)}$ and $\hat{P}^{f,F}_{\hat{\Lambda},(p_e)}$, given by (3.24), the FKG inequalities, (3.20), are valid. If $p'_e \le p_e$ for all e, then*

$$\hat{P}^{w,F}_{\hat{\Lambda}^*,(p'_e)} \quad \ll \quad \hat{P}^{w,F}_{\hat{\Lambda}^*,(p_e)},$$

$$\hat{P}^{f,F}_{\hat{\Lambda},(p'_e)} \quad \ll \quad \hat{P}^{f,F}_{\hat{\Lambda},(p_e)}.$$

(3.28)

The FK measure (with general coupling signs), $\hat{P}^{\bar{s}}_{\hat{\Lambda}^*,(\beta J_e)}$, given by (3.14), satisfies

$$\hat{P}^{\bar{s}}_{\hat{\Lambda}^*,(\beta J_e)} \quad \ll \quad \hat{P}^{w,F}_{\hat{\Lambda}^*,(p_e)}.$$

(3.29)

Proof. We give the arguments for wired b.c.'s; they are the same for free b.c.'s. If the wired ferromagnetic FK measure satisfies the FKG lattice condition, (3.21), then the FKG inequalities follow. To verify the lattice condition, we first note that the ratio of the LHS to the RHS of (3.21) can be written as a product of similar ratios, but where the two percolation configurations agree except for two edges e and e' where one configuration is $(1,0)$ and the other is $(0,1)$. To show that the ratio for all such situations is ≥ 1 is equivalent to showing that for any configuration n, the ratio

$$g\left((n_e : e \neq e')\right) \equiv \frac{\hat{P}^{w,F}_{\hat{\Lambda}^*,(p_e)}\left(n^{(e',1)}\right)}{\hat{P}^{w,F}_{\hat{\Lambda}^*,(p_e)}\left(n^{(e',0)}\right)} \quad \text{is increasing,} \tag{3.30}$$

where $n^{(e',\delta)}$ denotes the configuration that is the same as n except that $n_{e'}$ is replaced by δ. But

$$g\left((n_e : e \neq e')\right) = 2^{\#^w(n^{(e',1)}) - \#^w(n^{(e',0)})} \cdot \frac{p'_e}{1 - p'_e}. \tag{3.31}$$

Thus, g increasing is equivalent to $\#^w(n^{(e',1)}) - \#^w(n^{(e',0)})$ increasing. This difference is either -1 or 0. We need only show that when n_e (for $e \neq e'$) changes from 0 to 1, this difference can not change from 0 to -1. But if the difference was zero (when n_e was 0), that means that already (when n_e was 0 and $n_{e'}$ was 0) either x' and y' were in the same n-cluster or else they both are n-connected to $\partial\Lambda$, but then that property of x' and y' also holds when n_e is changed to 1 (and $n_{e'}$ is still 0) so the difference will still be zero (when n_e is changed to 1).

Now that the FKG inequalities have been validated, we obtain (3.28) and (3.29), providing the following are decreasing:

$$d\hat{P}^{w,F}_{\hat{\Lambda}^*,(p'_e)} \,/\, d\hat{P}^{w,F}_{\hat{\Lambda}^*,(p_e)} = \text{const.} \prod_{e \in \hat{\Lambda}^*} \left[\frac{1 - p_e}{p_e} \cdot \frac{p'_e}{1 - p'_e} \right]^{n_e} \tag{3.32}$$

and

$$d\hat{P}^{\bar{s}}_{\hat{\Lambda}^*,(\beta J_e)} \,/\, d\hat{P}^{w,F}_{\hat{\Lambda}^*,(p_e)} = \text{const.} \, 1_{U^{\bar{s}}}(n). \tag{3.33}$$

The first is decreasing because $p'_e(1 - p'_e)^{-1} \leq p_e(1 - p_e)^{-1}$. To see that the second is decreasing, note that the conditions for n to be in $U^{\bar{s}}$ are simply conditions that rule out certain n-loops (those with odd numbers of negative couplings) and n-paths (those connecting boundary sites where the relative sign of the b.c. spins disagrees with the coupling-sign parity of the path). Clearly the set of conditions only increases (or stays the same) when n increases and so $1_{U^{\bar{s}}}$ is decreasing. $\square$

The following proposition is well known, but the proof given here, based on the FK representation and the above inequalities, is nonstandard. (Usually the proof is based on other inequalities [Gr67], [KeS68], [ForKG71].) We write $E^{\pm,F}_{\Lambda,\beta}(S_A)$ and $E^{f,F}_{\Lambda,\beta}(S_A)$ to denote the expectations (w.r.t. the corresponding $+$, $-$ or free b.c. Gibbs measures) in a ferromagnetic Ising model of $S_A = \prod_{x \in A} S_x$. We write $P_\Lambda \to P$ as $\Lambda \to \mathbb{Z}^d$, where P_Λ (resp., P) is a probability measure on $\mathcal{S}_\Lambda$ (resp., $\mathcal{S}$), if for every sequence $\tilde{\Lambda}_n \subset \mathbb{Z}^d$ (with $\tilde{\Lambda}_n$ finite and such that $\forall L$, $\tilde{\Lambda}_n \supseteq \Lambda_L$ for all large n), $P_{\tilde{\Lambda}_n} \to P$ (in the sense of convergence of finite dimensional distributions).

Proposition 3.7 *If $A \subseteq \Lambda \subseteq \Lambda'$ are finite subsets of $\mathbb{Z}^d$, then*

$$E_{\Lambda,\beta}^{+,F}(S_A) \geq E_{\Lambda',\beta}^{+,F}(S_A) \geq 0, \tag{3.34}$$

$$0 \leq E_{\Lambda,\beta}^{f,F}(S_A) \leq E_{\Lambda',\beta}^{f,F}(S_A). \tag{3.35}$$

Hence, there exist infinite volume Gibbs states $P_\beta^{\pm,F}$ and $P_\beta^{f,F}$ such that

$$\text{for } b \in \{+,-,f\}, \quad P_{\Lambda,\beta}^{b,F} \to P_\beta^{b,F} \quad \text{as } \Lambda \to \mathbb{Z}^d. \tag{3.36}$$

Proof. We first note that Propositions 3.2 and 3.3 allow one to express the Gibbs expectations of S_A in terms of FK probabilities. These involve the n-partition of A, by which we mean the partition of A into (maximal) subsets $A_1, \ldots, A_m$ of sites which belong to the same n-cluster. The resulting expressions, which generalize (3.9) and (3.17), are (for $A \subseteq \Lambda$):

$$E_{\Lambda,\beta}^{f,F}(S_A) = \hat{P}_{\hat{\Lambda},\beta}^{f,F} \left(\text{ each } A_i \text{ in the } n\text{-partition of } A \text{ is even } \right), \tag{3.37}$$

$$E_{\Lambda,\beta}^{\pm,F}(S_A) = (\pm 1)^{|\Lambda|} \hat{P}_{\hat{\Lambda}^*,\beta}^{w,F} \left(\text{ each } A_i \text{ in the } n\text{-partition of } A, \tag{3.38} \right.$$

$$\text{which does not touch } \partial\Lambda, \text{ is even }).$$

These identities yield the non-negativity inequalities of (3.34)-(3.35). An elementary but key observation is that the two events appearing in (3.37)-(3.38) are both increasing (i.e., their indicator functions are increasing). Thus, by (3.28), for $b = +$ or f, $E_{\Lambda',\beta}^{b,F}(S_A)$ is a nondecreasing function of the p_e's (or equivalently of the J_e's). From the original definitions of the finite volume Gibbs distributions, it is easy to see that if we begin with $E_{\Lambda',\beta}^{w,F}(S_A)$, with $A \subseteq \Lambda$, then the limit, as $p_e \to 1$ ($J_e \to +\infty$) for every $e \in \hat{\Lambda}'^* \setminus \hat{\Lambda}^*$, is $E_{\Lambda,\beta}^{w,F}(S_A)$; this yields the rest of (3.34). Similarly, $E_{\Lambda',\beta}^{f,F}(S_A)$, in the limit, as $p_e \to 0$ for every $e \in \hat{\Lambda}' \setminus \hat{\Lambda}$, is $E_{\Lambda,\beta}^{f,F}(S_A)$; this yields the rest of (3.35). Monotonicity in Λ (for $b = w$ or f) implies well-defined limits as $\Lambda \to \mathbb{Z}^d$ of $E_{\Lambda,\beta}^{b,F}(S_A)$ for every A, which, by standard arguments, yield the final claim of the proposition. $\square$

The next proposition is also well known [LM72], but is usually proved by different methods. Part of the proof we give here may be regarded as a simplified version of the arguments we give later to obtain uniqueness of the infinite volume Gibbs state for nonferromagnetic models.

Proposition 3.8 *For ferromagnetic couplings, there is a unique infinite volume Gibbs state at inverse temperature β if and only if*

$$\forall x \in \mathbb{Z}^d, \quad E_\beta^{+,F}(S_x) \left(= \lim_{\Lambda \to \mathbb{Z}^d} \hat{P}_{\hat{\Lambda}^*,\beta}^{w,F}(x \overset{n}{\longleftrightarrow} \partial\Lambda) \right) = 0. \tag{3.39}$$

Proof. As β will be fixed throughout this proof, we drop the β subscript. The only if part of the propositon is simply due to the fact that the infinite volume Gibbs states $P^{+,F}$ and $P^{-,F}$ differ if $E^{+,F}(S_x) \neq 0$ for some x, since $E^{-,F}(S_x) = -E^{+,F}(S_x)$. The if part is more interesting. We will show that if

$$\forall x \in \mathbb{Z}^d, \quad \hat{P}_{\hat{\Lambda}^*}^{w,F}(x \overset{n}{\longleftrightarrow} \partial\Lambda) \to 0 \quad \text{as } \Lambda \to \mathbb{Z}^d, \tag{3.40}$$

then for any sequence of b.c.'s $\bar{s}^{(L)}$ on $\partial\Lambda_L$,

$$\forall \text{ finite } A \subseteq \mathbb{Z}^d, \quad E_{\Lambda_L}^{\bar{s}^{(L)},F}(S_A) \to E^{f,F}(S_A) \quad \text{as } L \to \infty. \tag{3.41}$$

Thus $P_{\Lambda_L}^{\bar{s}^{(L)},F} \to P^{f,F}$ (on $\mathcal{S}$) regardless of the choice of $\bar{s}^{(L)}$; by Proposition 3.1, this implies uniqueness, as desired.

To obtain (3.41), we will use Proposition 3.3 to derive an expression for the Gibbs expectation of S_A with b.c. $\bar{s}$ on $\partial\Lambda$ as a mixture of free b.c. expectations on various subsets $\tilde{\Lambda}$ plus a residual term. On the space $\mathcal{S}_{\Lambda^*} \times \mathcal{N}_{\hat{\Lambda}^*}$ (with probability measure $\tilde{P}^{\bar{s}} = \tilde{P}_{\Lambda^*,\hat{\Lambda}^*}^{\bar{s}}$), we denote by $C = C(\partial\Lambda)$ the n-cluster of $\partial\Lambda$, i.e.,

$$C = \left\{ x \in \Lambda : \ x \overset{n}{\longleftrightarrow} \partial\Lambda \right\}. \tag{3.42}$$

The key point is that (roughly speaking) if C is not all of Λ, then inside $\Lambda \setminus C$, $\tilde{P}^{\bar{s},F}$ is the same as $\tilde{P}^{f,F}$. More precisely and more specifically, for any (deterministic) nonempty $\tilde{\Lambda} \subseteq \Lambda$, the conditional distribution of $\{S_x : x \in \tilde{\Lambda}\}$, conditioned on the event that $\Lambda \setminus C = \tilde{\Lambda}$, is exactly $P_{\tilde{\Lambda}}^{f,F}$. (See the first few paragraphs of the proof of Theorem 3.11 below for a detailed discussion related to this point.) This immediately implies that for $A \subseteq A' \subseteq \Lambda$,

$$E_\Lambda^{\bar{s},F}(S_A) = \sum_{\Lambda \supseteq \tilde{\Lambda} \supseteq A'} \hat{P}_{\hat{\Lambda}^*}^{\bar{s},F}\left(\Lambda \setminus C = \tilde{\Lambda}\right) E_{\tilde{\Lambda}}^{f,F}(S_A)$$

$$+ \hat{P}_{\hat{\Lambda}^*}^{\bar{s},F}\left(A' \overset{n}{\longleftrightarrow} \partial\Lambda\right) \tilde{E}_{\Lambda^*,\hat{\Lambda}^*}^{\bar{s},F}\left(S_A \,\Big|\, A' \overset{n}{\longleftrightarrow} \partial\Lambda\right). \tag{3.43}$$

The remainder of the proof is the demonstration that (3.40) and (3.43) together imply (3.41). For a given A and any $\varepsilon > 0$, Proposition 3.7 implies that we may choose $A' = \Lambda_L$ in (3.43) with L so big that

$$\left| E_{\tilde{\Lambda}}^{f,F}(S_A) - E^{f,F}(S_A) \right| \le \varepsilon \quad \forall \tilde{\Lambda} \supseteq \Lambda_L. \tag{3.44}$$

Then for $\Lambda \supseteq \Lambda_L$, we have from (3.43) by elementary arguments that

$$\left| E_\Lambda^{\bar{s},F}(S_A) - E^{f,F}(S_A) \right| = \left| E_\Lambda^{\bar{s},F}(S_A) - \sum_{\Lambda \supseteq \tilde{\Lambda} \supseteq \Lambda_L} \hat{P}_{\hat{\Lambda}^*}^{\bar{s},F}\left(\Lambda \setminus C = \tilde{\Lambda}\right) E^{f,F}(S_A) \right.$$

$$\left. - \hat{P}_{\hat{\Lambda}^*}^{\bar{s},F}\left(\Lambda_L \overset{n}{\longleftrightarrow} \partial\Lambda\right) E^{f,F}(S_A) \right| \tag{3.45}$$

$$\le \ \varepsilon + 2\hat{P}_{\hat{\Lambda}^*}^{\bar{s},F}\left(\Lambda_L \overset{n}{\longleftrightarrow} \partial\Lambda\right).$$

But

$$\hat{P}_{\hat{\Lambda}^*}^{\bar{s},F}\left(\Lambda_L \overset{n}{\longleftrightarrow} \partial\Lambda\right) \le \sum_{x \in \Lambda_L} \hat{P}_{\hat{\Lambda}^*}^{\bar{s},F}\left(x \overset{n}{\longleftrightarrow} \partial\Lambda\right), \tag{3.46}$$

which, by (3.40), converges to zero as $\Lambda \to \mathbb{Z}^d$, for any fixed L. Thus, for any $\varepsilon > 0$,

$$\limsup_{\Lambda \to \mathbb{Z}^d} \left| E_\Lambda^{\bar{s},F}(S_A) - E^{f,F}(S_A) \right| \le \varepsilon, \tag{3.47}$$

which yields (3.41), as desired. $\qquad\square$

For a given set of couplings (with general signs), we define the critical inverse temperature

$$\beta_c = \sup\{\beta' \ge 0: \quad \text{there is a unique infinite volume} \tag{3.48}$$
$$\text{Gibbs state for every } \beta \in [0, \beta']\}.$$

The next proposition gives two useful (known) facts about β_c, one for ferromagnets and one for disordered systems.

Proposition 3.9 *For ferromagnetic couplings, there is non-uniqueness of infinite volume Gibbs states at any inverse temperature $\beta > \beta_c$. For disordered models (with general coupling signs), β_c is a.s. a constant.*

Proof. From the definition of β_c, if $\beta > \beta_c$, then there exists some $\beta' < \beta$ such that there is non-uniqueness of infinite volume Gibbs states at inverse temperature β'. Then, by Proposition 3.8, $E_{\beta'}^{+,F}(S_{x'}) \ne 0$ for some x', which implies, by (3.28) and (3.38), that $E_\beta^{+,F}(S_{x'}) > 0$. Applying Proposition 3.8 again, we obtain the first claim of the proposition (non-uniqueness at β).

To justify the second claim of the proposition, we begin by showing that β_c, which is a function of the J_e's and hence of ω, is measurable. For any finite $\Lambda \subset \mathbb{Z}^d$, any $\beta \in [0, \infty)$, any b.c. $\bar{s} \in \mathcal{S}_{\partial\Lambda}$ and any $A \subseteq \Lambda$, the Gibbs expectation, $E_{\Lambda,\beta}^{\bar{s}}(S_A)$, is an explicit rational function of $\{e^{\beta J_e} : e \in \hat{\Lambda}^*\}$ and hence is measurable. Also, since the denominator of the rational function does not vanish, $E_{\Lambda,\beta}^{\bar{s}}(S_A)$ is continuous in β. Thus

$$\delta_{\Lambda,\beta,A} \equiv \max_{\bar{s} \in \mathcal{S}_{\partial\Lambda}} E_{\Lambda,\beta}^{\bar{s}}(S_A) - \min_{\bar{s} \in \mathcal{S}_{\partial\Lambda}} E_{\Lambda,\beta}^{\bar{s}}(S_A) \tag{3.49}$$

is also measurable in the J_e's and continuous in β, and so the following functions of the J_e's are all measurable:

$$\Delta_{\Lambda,\beta',A} \equiv \sup_{\beta \in [0,\beta']} \delta_{\Lambda,\beta,A} = \sup_{\beta \in [0,\beta'] \cap \mathbb{Q}} \delta_{\Lambda,\beta,A}, \tag{3.50}$$

$$\Delta_{\beta',A} \equiv \lim_{L \to \infty} \sup_{\substack{\Lambda \ finite \\ \Lambda \supseteq \Lambda_L}} \Delta_{\Lambda,\beta',A}, \tag{3.51}$$

$$\Delta_{\beta'} \equiv \sup_{A \ finite} \Delta_{\beta',A}, \tag{3.52}$$

$$\Delta \equiv \sup_{\beta' \in [0,\infty) \cap \mathbb{Q}} \beta' 1_{\Delta_{\beta'}=0}. \tag{3.53}$$

We claim that Δ is identical to the function β_c defined in (3.48). To see this, first note that if $\beta'' < \Delta$, then $\exists \beta' > \beta''$ such that $\Delta_{\beta'} = 0$. But any two infinite volume Gibbs states, P_β^1 and P_β^2, with $\beta \le \beta'$, satisfy

$$\left| E_\beta^1(S_A) - E_\beta^2(S_A) \right| \le \Delta_{\beta',A} \le \Delta_{\beta'}. \tag{3.54}$$

Thus $\Delta_{\beta'} = 0$ implies $E_\beta^1(S_A) = E_\beta^2(S_A)$ for all A and thus that $P_\beta^1 = P_\beta^2$; i.e., $\Delta_{\beta'} = 0$ implies uniqueness for all $\beta \le \beta'$. This shows that $\beta' \le \beta_c$ and thus that $\beta'' < \Delta$ implies $\beta'' < \beta_c$; i.e., $\Delta \le \beta_c$.

To complete the demonstration that $\Delta = \beta_c$, and thus that β_c is measurable, it remains to show that $\Delta \ge \beta_c$. So suppose $\beta'' > \Delta$; we must show that $\beta'' > \beta_c$, i.e., that for some $\beta' < \beta''$, there is non-uniqueness (of infinite volume Gibbs states). But $\beta'' > \Delta$ implies that $\exists \bar{\beta} \in (0, \beta'')$ such that $\Delta_{\bar{\beta}} \neq 0$; we will show that $\Delta_{\bar{\beta}} \neq 0$ implies non-uniqueness for some $\beta' < \bar{\beta}$. To show this, we first note that $\Delta_{\bar{\beta}} \neq 0$ implies that for some A, some sequence $\Lambda^{(L)} \supseteq \Lambda_L$, some sequence $\beta_L \in [0, \bar{\beta}]$ and some b.c.'s $\bar{s}^{(L)}$ and $\bar{t}^{(L)}$ in $\mathcal{S}_{\partial \Lambda^{(L)}}$,

$$E_{\Lambda^{(L)}, \beta_L}^{\bar{s}^{(L)}}(S_A) - E_{\Lambda^{(L)}, \beta_L}^{\bar{t}^{(L)}}(S_A) \to \Delta_{\bar{\beta}, A} > 0. \tag{3.55}$$

Furthermore, by using standard compactness arguments and choosing a subsequence, we may assume that $\beta_L \to \beta' \in [0, \bar{\beta}]$, and that $P_{\Lambda^{(L)}, \beta_L}^{\bar{s}^{(L)}}$ and $P_{\Lambda^{(L)}, \beta_L}^{\bar{t}^{(L)}}$ converge to probability measures P^1 and P^2 on $\mathcal{S}$. It is not hard to see, by using the DLR equations, that P^1 and P^2 are infinite volume Gibbs measures at inverse temperature β'. But they are distinct since by (3.55), their expectations of S_A differ. Thus there is non-uniqueness at some $\beta' < \bar{\beta}$, which completes the demonstration that β_c is measurable.

Now that we know β_c is measurable w.r.t. $\mathcal{F}$, we will show that it is actually measurable with respect to the tail field, i.e., it does not depend on any finite collection of J_e's. Then the second claim of the proposition will follow from Kolmogorov's zero-one law. (One can instead, and probably more easily, use translation invariance of β_c and translation ergodicity of the J_e's to obtain the second claim of the proposition, but the fact that β_c is a tail variable is of some interest itself, so we use that approach.)

Consider a Gibbs state $P_{\Lambda, \beta}^{\bar{s}}$ for a given set of J_e's and a Gibbs state $P_{\Lambda, \beta}^{\bar{s}, *}$ for couplings J_e^* which differ from J_e on a fixed set $\hat{B}$ of edges with $\hat{B} \subseteq \hat{\Lambda}$. Then one has the identities,

$$\forall A \subseteq \Lambda, \quad E_{\Lambda, \beta}^{\bar{s}, *}(S_A) = \frac{E_{\Lambda, \beta}^{\bar{s}}(G^* S_A)}{E_{\Lambda, \beta}^{\bar{s}}(G^*)}, \tag{3.56}$$

where

$$G^* = \exp\left(\beta \sum_{\{x,y\} \in \hat{B}} (J_e^* - J_e) S_x S_y \right). \tag{3.57}$$

These identities and the corresponding ones with the J_e's and J_e^*'s interchanged, combined with Proposition 3.1, show that uniqueness for the J_e's at a given β is equivalent to uniqueness for the J_e^*'s at that β. Thus β_c is the same for the J_e's and J_e^*'s which shows that β_c is a tail variable, as claimed. This completes the proof. □

According to Proposition 3.9, for disordered systems, a.s.

$$\beta_c = \sup\{\beta' \geq 0: \quad \text{a.s., there is a unique infinite volume} \qquad (3.58)$$
$$\text{Gibbs state for every } \beta \in [0, \beta']\}.$$

Henceforth, we treat β_c as this constant value. An alternative definition of a critical value is

$$\bar{\beta}_c = \sup\{\beta \geq 0: \quad \text{a.s, there is a unique infinite} \qquad (3.59)$$
$$\text{volume Gibbs state for } \beta\}.$$

For ferromagnetic couplings, Proposition 3.9 shows that $\beta_c = \bar{\beta}_c$. In general, one only knows that $\beta_c \leq \bar{\beta}_c$ and, if the first part of Proposition 3.9 does not extend to models with general coupling signs, it is conceivable that for some models, $\beta_c < \bar{\beta}_c$ or even that $\beta_c < \infty$ while $\bar{\beta}_c = \infty$!

We are now ready to present our first type of result which gives upper and lower bounds on β_c for ferromagnetic disordered systems. These bounds are in terms of the critical value $p_c = p_c(d)$ for standard homogeneous Bernoulli bond percolation on $(\mathbb{Z}^d, \mathbb{E}^d)$. This critical value may be defined as follows. Let $\hat{P}^{ind}_{\mathbb{E}^d,(p)}$ be the product measure on $\{0,1\}^{\mathbb{E}^d}$ corresponding to independent n_e's with $\hat{P}^{ind}_{\mathbb{E}^d,(p)}(n_e = 1) = p$ for each $e \in \mathbb{E}^d$. Let $x \overset{n}{\longleftrightarrow} \infty$ denote the event that the n-cluster of x is infinite. The first main result of percolation theory [BroH57], [Ha57], [Ha59] is that for $d \geq 2$, there is a $p_c \in (0,1)$ such that

$$\theta_d(p) \equiv \hat{P}^{ind}_{\mathbb{E}^d,(p)}(x \overset{n}{\longleftrightarrow} \infty) \begin{cases} = 0 & \text{for } p < p_c, \\ > 0 & \text{for } p > p_c. \end{cases} \qquad (3.60)$$

(See the related discussion in Chapter 2 between Theorems 2.5 and 2.6.)

Theorem 3.10 ([ACCN87]) *For a disordered ferromagnet with $d \geq 2$ and a given common distribution μ on $[0,\infty)$ for the J_e's, define*

$$\bar{p}(\beta) = E(p_e), \qquad \underline{p}(\beta) = E(h(p_e)), \qquad (3.61)$$

where p_e is given by (3.2) and $h(p_e)$ by (3.26). Then for a given β, there is a.s. uniqueness (resp., non-uniqueness) of infinite volume Gibbs states if $\bar{p}(\beta) < p_c$ (resp., $\underline{p}(\beta) > p_c$). It follows that $\beta_c > 0$ for any μ and that

$$\beta_c < \infty \iff \mu(\{0\}) < 1 - p_c, \qquad (3.62)$$

and (when $\beta_c < \infty$) that $\beta^l \leq \beta_c \leq \beta^u$ where β^l (resp., β^u) is the (unique) solution of $\bar{p}(\beta) = p_c$ (resp., $\underline{p}(\beta) = p_c$).

Proof. According to Proposition 3.8, to obtain uniqueness (resp., non-unique-ness), we need to show that (3.39) is a.s. valid (resp., not valid) when $\bar{p}(\beta) < p_c$ (resp., $\underline{p}(\beta) > p_c$). But, from (3.25),

$$\lim_{\Lambda \to \mathbb{Z}^d} \hat{P}^{w,F}_{\Lambda^*,\beta}(x \xleftrightarrow{n} \partial\Lambda) \begin{cases} \leq \lim_{\Lambda \to \mathbb{Z}^d} \hat{P}^{ind}_{\Lambda^*,(p_e)}(x \xleftrightarrow{n} \partial\Lambda) = \hat{P}^{ind}_{\mathbb{E}^d,(p_e)}(x \xleftrightarrow{n} \infty), \\ \geq \lim_{\Lambda \to \mathbb{Z}^d} \hat{P}^{ind}_{\Lambda^*,(h(p_e))}(x \xleftrightarrow{n} \partial\Lambda) = \hat{P}^{ind}_{\mathbb{E}^d,(h(p_e))}(x \xleftrightarrow{n} \infty). \end{cases}$$

$$(3.63)$$

Thus, it suffices to show that for $\hat{P}^{ind}_{\mathbb{E}^d,(p'_e)}$, a disordered percolation measure with i.i.d. p'_e's, one has:

$$\text{a.s.,} \quad \hat{P}^{ind}_{\mathbb{E}^d,(p'_e)}(\text{ some } n\text{-cluster is infinite}) = \begin{cases} 0, & \text{if } E(p'_e) < p_c, \\ 1, & \text{if } E(p'_e) > p_c. \end{cases} \quad (3.64)$$

Letting E denote expectation on the probability space of the p'_e's, me make the crucial, albeit essentially trivial, observation that

$$E\left(\hat{P}^{ind}_{\mathbb{E}^d,(p'_e)}\right) = \hat{P}^{ind}_{\mathbb{E}^d,(E(p'_e))}. \quad (3.65)$$

This is because the p'_e's are independent and, conditionally on the p'_e's, the n_e's are independent, which implies that the n_e's are (marginally) independent; also, the marginal probability that $n_e = 1$ is clearly $E(p'_e)$. Now by definition (3.60) of p_c and ergodicity, we have

$$\hat{P}^{ind}_{\mathbb{E}^d,(E(p'_e))}(\text{ some } n\text{-cluster is infinite}) = \begin{cases} 0, & \text{if } E(p'_e) < p_c, \\ 1, & \text{if } E(p'_e) > p_c, \end{cases} \quad (3.66)$$

which, together with (3.65), immediately implies (3.64). This proves the first set of claims of the proposition.

To see that $\beta_c > 0$ for any μ, note that

$$\bar{p}(\beta) = E\left(1 - e^{-2\beta J_e}\right) \to 0 \quad \text{as } \beta \to 0^+, \quad (3.67)$$

so $\bar{p}(\beta) < p_c$ for small β, which implies a.s. uniqueness for small β and thus that $\beta_c > 0$. To prove (3.62), we first note that if $\mu(\{0\}) \geq 1 - p_c$, then for any $\beta < \infty$,

$$\bar{p}(\beta) = E\left(1 - e^{-2\beta J_e}\right) < \nu(J_e > 0) = 1 - \mu(\{0\}) \leq p_c. \quad (3.68)$$

Thus there is a.s. uniqueness for any $\beta < \infty$, so $\beta_c = \infty$. On the other hand, if $\mu(\{0\}) < 1 - p_c$, then, since

$$h(p) = \frac{p}{2(1-p) + p} \to 1 \quad \text{as } p \to 1^-, \quad (3.69)$$

while $h(0) = 0$, we see that

$$\underline{p}(\beta) = E\left(h(1 - e^{-2\beta J_e})\right) \to \nu(J_e > 0) = 1 - \mu(\{0\}) > p_c \quad \text{as } \beta \to \infty. \quad (3.70)$$

Thus there is a.s. non-uniqueness for large enough β, so $\beta_c < \infty$. This proves (3.62) and the bounds $\beta^l \leq \beta_c \leq \beta^u$ follow from similar considerations. $\qquad \square$

Proposition 3.9 and Theorem 3.10 give fairly good control on the uniqueness and non-uniqueness regimes for disordered ferromagnets. For other disordered systems, such as spin glasses, the tools and results are much less developed. There are no rigorous results giving non-uniqueness for spin glasses, but there are many open problems and conjectures, some of which will be discussed in the next chapter. There do exist rigorous results giving uniqueness for high temperatures, such as those of [Do68], [DoS85] (which require some restrictions on μ) and those of [BD86], [DrKP95], [GiM95] (which do not). The last main result of this chapter is a uniqueness result, taken from [N94], which also has no restrictions on μ but gives a better lower bound on β_c. For a model with general coupling signs and critical value β_c, we denote by β_c^F the critical value for the associated ferromagnet in which each J_e is replaced by $|J_e|$.

Theorem 3.11 ([N94]) *There is uniqueness of the infinite volume Gibbs state at inverse temperature β for a model with couplings $\{J_e : e \in \mathbb{E}^d\}$ of general sign if there is uniqueness at the same β for the associated ferromagnet with couplings $\{|J_e| : e \in \mathbb{E}^d\}$. Thus for a general model with arbitrary μ,*

$$\beta_c \geq \beta_c^F \geq \beta^l > 0. \tag{3.71}$$

Here, β^l is the solution, β, of the equation

$$E\left(1 - e^{-2\beta|J_e|}\right) = p_c(d) \tag{3.72}$$

when $P(|J_e| \neq 0) > p_c(d)$, and $\beta^l = \infty$ when $P(|J_e| \neq 0) \leq p_c(d)$.

Proof. The second part of the theorem is an immediate consequence of the first part and Theorem 3.10. The proof of the first part combines the FK representation (including the ferromagnetic criterion (3.39) for uniqueness) with extensions of the comparison inequality (3.29) and a coupling between nonferromagnets and the associated ferromagnets. The coupling argument is similar in spirit to one used in [BM94].

We begin by extending (3.29) to more general "boundary" conditions. (A similar analysis of general FK boundary conditions may be found in [Pi96].) The b.c.'s that we must handle are the kind that implicitly arise when we begin with $\hat{P}^{\bar{s}}_{\hat{\Lambda}^*,(\beta J_e)}$ (or more generally $\tilde{P}^{\bar{s}}_{\hat{\Lambda}^*,\hat{\Lambda}^*,(\beta J_e)}$), and look at the conditional distribution of $n' = (n_e : e \in \hat{\Lambda}' \subseteq \hat{\Lambda}^*)$, given $n^c = (n_e : e \in \hat{\Lambda}^* \setminus \hat{\Lambda}')$. We remark, as pointed out by A. Pisztora, that a detailed analysis of b.c.'s can be avoided by relying on the fact that such conditioning preserves the FKG lattice condition (3.21). Nevertheless, we continue with our b.c. considerations, since they may be useful in other situations.

For a given n^c, $\Lambda^* = \Lambda \cup \partial\Lambda$ is partitioned into $\Lambda_1, \ldots, \Lambda_m$, where

$$\Lambda_1 = \partial\Lambda \cup \{x \in \Lambda : x \xleftrightarrow{n^c} \partial\Lambda\} \tag{3.73}$$

and $\Lambda_2, \ldots, \Lambda_m$ are the vertex-sets of the n^c-clusters which do not touch $\partial \Lambda$. For each $x \in \Lambda_1$, the spin s_x is determined by $\bar{s}$, n^c and the coupling signs, so that $(s, n) \in \tilde{U}^{\bar{s}}_{\Lambda^*, \hat{\Lambda}^*}$ (see (3.13)). Similarly for any j and $x, y \in \Lambda_j$, the relative sign $s_x s_y$ is determined. Since $\bar{s}$ and $-\bar{s}$ yield the same FK measure, we may put Λ_1 on an equal footing with the other Λ_i's by considering only the relative signs also for the sites in Λ_1. Thus we regard our conditional distribution $\hat{P}^\psi_{\hat{\Lambda}', (\beta J_e)}$ on $\mathcal{N}_{\hat{\Lambda}'}$ as having a boundary condition ψ which is a partition $\Lambda_1, \ldots, \Lambda_m$ of all the sites in Λ^* together with an assignment t_i of relative signs to each Λ_i. The formula for this distribution is

$$\hat{P}^\psi_{\hat{\Lambda}', (\beta J_e)}(\{n'\}) = \left(\hat{Z}^\psi_{\hat{\Lambda}', (\beta J_e)} \right)^{-1} 1_{U^\psi}(n') \, 2^{\#^\psi(n')} \hat{P}^{ind}_{\hat{\Lambda}', (p_e)}(\{n'\}), \qquad (3.74)$$

where

$$\begin{aligned} U^\psi \;\; = \;\; & \{n' \in \mathcal{N}_{\hat{\Lambda}'} : \exists s \in \mathcal{S}_\Lambda \text{ with } s_x s_y \text{ given by } t_i \text{ whenever } x \text{ and} \\ & y \text{ are in the same } \Lambda_i \text{ and such that } \forall e = \{x, y\} \in \hat{\Lambda}', \qquad (3.75) \\ & n'_e J_e s_x s_y \geq 0\}, \end{aligned}$$

and $\#^\psi(n')$ is (roughly speaking) the number of n'-clusters when the vertices in each Λ_i are identified (or regarded as already connected). More precisely, $\#^\psi(n')$ is the number of connected components in the graph with vertex set Λ^* and edge set

$$\left\{ e \in \hat{\Lambda}' : n'_e = 1 \right\} \cup \{\{x, y\} : \text{ for some } i, \; x \in \Lambda_i \text{ and } y \in \Lambda_i \text{ and } x \neq y\}; \quad (3.76)$$

in this definition we include $\{x, y\}$'s not in $\mathbb{E}^d$ when x and y are in the same Λ_i.

Before giving the extension of (3.29), we discuss a few instructive examples of such boundary conditions. The free b.c. FK measure, $\hat{P}^f_{\hat{B}, (\beta J_e)}$ (see (3.5)), for a finite $B \subset \mathbb{Z}^d$, equals the measure of (3.74) providing $\hat{\Lambda}'$ (a subset of $\hat{\Lambda}$) and $\{\Lambda_i\}$ (a partition of Λ^* which is part of the b.c. ψ) are such that $\hat{\Lambda}' = \hat{B}$ and no two sites of B are in the same Λ_i. One way this occurs, as in the derivation of (3.43) above, is when one begins with $\hat{P}^{\bar{s}}_{\hat{\Lambda}^*, (\beta J_e)}$ and conditions on an $n^c = (n_e : e \in \hat{\Lambda}^* \setminus \hat{B})$ such that $n_e = 0$ for every $e \in \hat{\Lambda}^* \setminus \hat{B}$ touching any site of B. Another example is the fixed $\bar{s}$ measure, $\hat{P}^{\bar{s}}_{\hat{B}^*, (\beta J_e)}$, (see (3.14)), which equals the measure of (3.74) providing $\hat{\Lambda}' = \hat{B}^*$, no two sites of B are in the same Λ_i, all the sites of ∂B are in the same Λ_j with that Λ_j disjoint from B and finally the relative sign assignment t_j for that Λ_j (restricted to ∂B) agrees with that given by $\bar{s}$. The simplest special case of this last example is when $B = \Lambda$, the partition $\{\Lambda_1, \ldots, \Lambda_m\}$ of Λ^* has $\Lambda_1 = \partial \Lambda$ with all other Λ_i's being single sites and t_1 is given by $\bar{s}$ (with no t_i's needed for $i \geq 2$). Partitions which are refinements of the one in this last case may be regarded as corresponding to b.c.'s intermediate between fixed $\bar{s}$ on the entire $\partial \Lambda$ and free. One such situation, giving our final examples, is when Λ is the cube Λ_L and we partition Λ_L into single sites and $\partial\Lambda_L$ into the pairs of sites in corresponding positions on opposing faces. When the t_j for every such pair $\Lambda_j = \{x, y\}$ is $s_x s_y = +1$ (resp., -1), these are periodic (resp., anti-periodic) b.c.'s.

The extension of (3.29) to our more general b.c.'s is as follows. For a given Λ^*, let ψ' be a b.c. consisting of $(\Lambda'_1, \ldots, \Lambda'_{m'})$ and $(t'_1, \ldots, t'_{m'})$ in which for each j, t'_j assigns the same sign to every s_x with $x \in \Lambda'_j$. For ferromagnetic couplings, note that $U^{\psi'}$ is all of $\mathcal{N}_{\hat{\Lambda}'}$ so that $1_{U^{\psi'}}$ in (3.74) may be deleted; this may be thought of as a "partially wired" b.c. Let ψ be any b.c. whose partition of Λ^* is a refinement of that of ψ', with any assignment of relative signs. Then (for any $\hat{\Lambda}'$)

$$\hat{P}^{\psi}_{\hat{\Lambda}',(\beta J_e)} \ll \hat{P}^{\psi',F}_{\hat{\Lambda}',(p_e)}. \tag{3.77}$$

As in the derivation of (3.29), this comparison inequality follows from two facts and an application of Proposition 3.4. The first fact is that the FKG inequalities (see (3.20)) are valid for the partially wired FK measure $\hat{P}^{\psi',F}_{\hat{\Lambda}',(p_e)}$. Just like for the usual wired measure $\hat{P}^{w,F}_{\hat{\Lambda}^*,(p_e)}$ (see the proof of Proposition 3.6), this is because, for any $e' \in \hat{\Lambda}'$, $\#^{\psi'}(n'^{(e',1)}) - \#^{\psi'}(n'^{(e',0)})$ is increasing (on $\mathcal{N}_{\hat{\Lambda}'}$). The second fact is that

$$d\hat{P}^{\psi}_{\hat{\Lambda}',(\beta J_e)} / d\hat{P}^{\psi',F}_{\hat{\Lambda}',(p_e)} = \text{const.} \, 1_{U^{\psi}}(n') \cdot 2^{\#^{\psi}(n') - \#^{\psi'}(n')}, \tag{3.78}$$

which is decreasing because both $1_{U^{\psi}}$ and $\#^{\psi} - \#^{\psi'}$ are decreasing. $1_{U^{\psi}}$ is decreasing because when n' increases, the set of conditions required for n' to be in U^{ψ} (see (3.75)) increases. $\#^{\psi} - \#^{\psi'}$ is decreasing because when a single n'_e is changed from 0 to 1, each of $\#^{\psi}$ and $\#^{\psi'}$ either stays the same or decreases by 1 but it is impossible for $\#^{\psi}$ to stay the same while $\#^{\psi'}$ decreases. This is due to the definitions of $\#^{\psi}$ (and similarly $\#^{\psi'}$) in terms of the graph on Λ^* with edge set given by (3.76) and the refinement relation between ψ and ψ' which implies that all the edges given by (3.76) for ψ are included in the edges given for ψ', so that if the edges x and y were in distinct components of the ψ'-graph before n'_e was changed (which is equivalent to $\#^{\psi'}$ decreasing because of this change), then they were also in distinct components of the ψ-graph (so that $\#^{\psi}$ also decreases because of this change). We have now proved (3.77).

We next construct a coupling between the nonferromagnetic $\tilde{P}^{\bar{s}} = \tilde{P}^{\bar{s}}_{\Lambda^*,\hat{\Lambda}^*,(\beta J_e)}$ with fixed b.c. $\bar{s}$ on $\partial\Lambda$ and the ferromagnetic $\tilde{P}^{w,F} = \tilde{P}^{w,F}_{\Lambda^*,\hat{\Lambda}^*,(p_e)}$ (defined as $\frac{1}{2}\tilde{P}^{+,F} + \frac{1}{2}\tilde{P}^{-,F}$) with wired b.c.; i.e., we construct a combined set of random variables $\{(S^{\bar{s}}_x, N^{\bar{s}}_e), (S^{w,F}_x, N^{w,F}_e) : x \in \Lambda^*, e \in \hat{\Lambda}^*\}$ on a single probability space so that the marginal distributions of the individual sets are $\tilde{P}^{\bar{s}}$ and $\tilde{P}^{w,F}$. Let us denote by $C^{w,F} = C^{w,F}(\partial\Lambda)$ the $N^{w,F}$-cluster of $\partial\Lambda$ and by $C^{\bar{s}}$ the $N^{\bar{s}}$-cluster of $\partial\Lambda$ (see (3.42)). Then by the properties of the marginal distributions of each of the individual sets of random variables, it follows (see the derivation of (3.43) and the discussion about boundary conditions at the beginning of this proof) that conditioned on $\Lambda \setminus C^{\bar{s}} = \tilde{\Lambda}$ (resp., $\Lambda \setminus C^{w,F} = \tilde{\Lambda}$) the conditional distribution of $(S_x^{\bar{s}} : x \in \tilde{\Lambda})$ (resp., $(S_x^{w,F} : x \in \tilde{\Lambda})$) is exactly $P_{\tilde{\Lambda},(\beta J_e)}^f$ (resp., $P_{\tilde{\Lambda},(\beta|J_e|)}^{f,F}$). Our particular coupling will be constructed so that something else, mixing together the two sets of variables, is valid:

$$\text{conditioned on } \Lambda \setminus C^{w,F} = \tilde{\Lambda}, \text{ the cond. dist. of} \atop (S_x^{\bar{s}} : x \in \tilde{\Lambda}) \text{ is exactly } P_{\tilde{\Lambda},(\beta J_e)}^f. \tag{3.79}$$

The main part of the coupling is a construction of the FK edge variables $\{N_e^{\bar{s}}, N_e^{w,F} : e \in \hat{\Lambda}^*\}$. We begin with a fixed deterministic ordering of all the edges in $\hat{\Lambda}^*$ and i.i.d. random variables $M_1, M_2, \ldots$, on some $(\Omega, \mathcal{F}, P)$, that are uniformly distributed on $[0,1]$. Later, when constructing the spin variables $\{S_x^{\bar{s}}, S_x^{w,F} : x \in \Lambda^*\}$, we will also need i.i.d., ± 1 valued, symmetric random variables $S_1, S_1', S_2, S_2', \ldots$, on the same $(\Omega, \mathcal{F}, P)$, that are independent of the M_i's. The construction is inductive, in that for each $j = 1, 2, \ldots, |\hat{\Lambda}^*|$, we will choose a certain edge $e_j = e_j(\omega) \in \hat{\Lambda}^*$, which is measurable w.r.t.

$$\mathcal{F}_{j-1} = \sigma(e_i, N_{e_i}^{\bar{s}}, N_{e_i}^{w,F} : i \leq j-1) = \sigma(M_i : i \leq j-1), \tag{3.80}$$

and then determine $N_{e_j}^{\bar{s}}, N_{e_j}^{w,F}$ by the value of M_j, according to a rule which is also $\mathcal{F}_{j-1}$-measurable. ($\mathcal{F}_0$ is of course the trivial σ-field.) The rule, which we give next, is fairly obvious (although it may look complicated); the choice of e_j, which we give a little later, is a bit more subtle.

Let $\hat{\Lambda}^j = \hat{\Lambda}^j(\omega)$ denote $\hat{\Lambda}^* \setminus \{e_1, \ldots, e_j\}$ and define the ω-dependent ($\mathcal{F}_j$-measurable) probability measure on $\mathcal{N}_{\hat{\Lambda}^j}$, $\hat{P}_j^{\bar{s}}$ (resp., $\hat{P}_j^{w,F}$), as the conditional distribution of $\hat{P}_{\hat{\Lambda}^*,(\beta J_e)}^{\bar{s}}$ (resp., $\hat{P}_{\hat{\Lambda}^*,(p_e)}^{w,F}$) on $\mathcal{N}_{\hat{\Lambda}^j}$, conditioned on $(n_{e_i} = N_{e_i}^{\bar{s}} : i \leq j)$ (resp., $(n_{e_i} = N_{e_i}^{w,F} : i \leq j)$). Then

$$N_{e_j}^{\bar{s}} \text{ (resp., } N_{e_j}^{w,F}) = \begin{cases} 1, & \text{if } M_j \leq \hat{P}_{j-1}^{\bar{s}}(n_{e_j} = 1) \text{ (resp., } \hat{P}_{j-1}^{w,F}(n_{e_j} = 1)), \\ 0, & \text{otherwise.} \end{cases}$$
$$\tag{3.81}$$

This rule, of course, guarantees that for each j, the conditional distribution of the single variable $N_{e_j}^{\bar{s}}$ (resp., $N_{e_j}^{w,F}$), given the previous ones, agrees with the corresponding conditional distribution from $\hat{P}_{\hat{\Lambda}^*,(\beta J_e)}^{\bar{s}}$ (resp., $\hat{P}_{\hat{\Lambda}^*,(p_e)}^{w,F}$). It follows that, regardless of our precise choice of the $\mathcal{F}_{j-1}$-measurable e_j, the marginal distribution of $(N_e^{\bar{s}} : e \in \hat{\Lambda}^*)$ (resp., $(N_e^{w,F} : e \in \hat{\Lambda}^*)$) is indeed $\hat{P}_{\hat{\Lambda}^*,(\beta,J_e)}^{\bar{s}}$ (resp., $\hat{P}_{\hat{\Lambda}^*,(p_e)}^{w,F}$); i.e., we do have a coupling. The reason for a careful choice of e_j is to insure that the coupling will satisfy (3.79).

We will choose the e_j's so that all the edges which touch the sites of $C^{w,F} = C^{w,F}(\partial\Lambda)$ are explored before any of the other edges of $\hat{\Lambda}^*$. The choice is given inductively as follows. Define $N^j = N^{j,w,F}$ in $\mathcal{N}_{\hat{\Lambda}^*}$ by

$$N_e^j = \begin{cases} N_{e_i}^{w,F}, & \text{if } e = e_i \text{ for } i \leq j-1, \\ 0, & \text{otherwise,} \end{cases} \tag{3.82}$$

and then $C^j = C^{j,w,F}$ and $\bar{C}^j$ by

$$C^j = \{x \in \Lambda : x \xrightarrow{N^j} \partial\Lambda\}, \qquad \bar{C}^j = C^j \cup \partial\Lambda. \tag{3.83}$$

e_1 is chosen as the edge in $\hat{\Lambda}^*$ touching $\partial\Lambda$ which is earliest in the deterministic ordering. For $j = 2, 3, \ldots$, e_j is chosen as that edge in

$$D_{j-1} \equiv \{e \in \hat{\Lambda}^* : e \text{ touches } \bar{C}^{j-1}\} \setminus \{e_1, \ldots, e_{j-1}\} \tag{3.84}$$

which is earliest in the deterministic ordering, providing D_{j-1} is nonempty. If J^* denotes the first j such that D_j is empty, then $C^{J^*} = C^{w,F}$. There are two ways this can happen. Either $J^* = |\hat{\Lambda}^*|$, in which case we have finished choosing all the e_j's or else $J^* < |\hat{\Lambda}^*|$. In the latter case, we simply choose e_j for $J^* < j \leq |\hat{\Lambda}^*|$ according to the deterministic ordering of the edges in $\hat{\Lambda}^{J^*} = \hat{\Lambda}^* \setminus \{e_1, \ldots, e_{J^*}\}$.

From the discussion about boundary conditions at the beginning of this proof, we can rewrite the conditional measure $\hat{P}^{\bar{s}}_{j-1}$ (resp., $\hat{P}^{w,F}_{j-1}$) appearing in (3.81) as the b.c. measure $\hat{P}^{\psi_{j-1}}_{\hat{\Lambda}^{j-1},(\beta J_e)}$ (resp., $\hat{P}^{\psi'_{j-1},F}_{\hat{\Lambda}^{j-1},(p_e)}$) for some b.c. ψ_{j-1} (resp., ψ'_{j-1}) determined by $(N^{\bar{s}}_{e_i} : i \leq j-1)$ and $\bar{s}$ (resp., by $(N^{w,F}_{e_i} : i \leq j-1)$). Note that ψ'_{j-1} is a partially wired measure (see the discussion before (3.77)) and that if

$$N^{\bar{s}}_{e_i} \leq N^{w,F}_{e_i} \quad \text{for } i \leq j-1, \tag{3.85}$$

then the partition of ψ_{j-1} is a refinement of the partition of ψ'_{j-1}, so that by (3.77) and the basic coupling step (3.81), it would follow that

$$N^{\bar{s}}_{e_j} \leq N^{w,F}_{e_j}. \tag{3.86}$$

This argument (including for the case $j = 1$) shows, by induction, that (3.86) is valid for all j. This in turn implies that, after step J^*, all the edges e between $C^{J^*} = C^{w,F}$ and $\Lambda \setminus C^{w,F}$ have $N^{w,F}_e = N^{\bar{s}}_e = 0$. Thus,

$$\hat{P}^{\bar{s}}_{J^*} = \hat{P}^f_{\hat{\Lambda}^{J^*},(\beta J_e)}, \qquad \hat{P}^{w,F}_{J^*} = \hat{P}^{f,F}_{\hat{\Lambda}^{J^*},(p_e)}. \tag{3.87}$$

Note also that $\hat{\Lambda}^{J^*}$ is just the set of edges $e = \{x, y\} \in \hat{\Lambda}^*$ such that x and y are in $\Lambda \setminus C^{w,F}$.

Once we have constructed all the $N^{\bar{s}}_e$'s and $N^{w,F}_e$'s, the spin variables $S^{\bar{s}}_x$ and $S^{w,F}_x$ are constructed, according to the coin-tossing rules of Proposition 3.3, using the "coins" $S_1, S_2, \ldots$ (one for each $N^{\bar{s}}$-cluster not touching $\partial\Lambda$ according to some reasonable ordering) to obtain the spins $S^{\bar{s}}_x$ and using the independent coins $S'_1, S'_2, \ldots$ to obtain the spins $S^{w,F}_x$. The LHS of (3.87) implies the crucial property (3.79) of our coupling.

The remainder of the proof proceeds along the lines of the last part of the proof of Proposition 3.8. Property (3.79) immediately implies, as an analogue of

(3.43), for $A \subseteq A' \subseteq \Lambda$

$$E_{\Lambda,(\beta J_e)}^{\bar{s}}(S_A) = \sum_{\Lambda \supseteq \tilde{\Lambda} \supseteq A'} \hat{P}_{\hat{\Lambda}^*,(p_e)}^{w,F}(\Lambda \setminus C^{w,F} = \tilde{\Lambda}) \, E_{\tilde{\Lambda},(\beta J_e)}^f(S_A)$$

$$+ \hat{P}_{\hat{\Lambda}^*,(p_e)}^{w,F}(A' \xleftrightarrow{n} \partial\Lambda) \, E(S_A^{\bar{s}} | A' \xleftrightarrow{N^{w,F}} \partial\Lambda). \quad (3.88)$$

Here, the final expectation, denoted simply E, is in the space $(\Omega, \mathcal{F}, P)$ of our coupling. If there is uniqueness of the infinite volume Gibbs state for the associated ferromagnet, then (3.39) is valid, which, together with (3.88), implies, as an analogue of (3.47), that for any two sequences of boundary conditions $\bar{s}^{(L)}$ and $\bar{t}^{(L)}$ on $\partial\Lambda_L$ and any finite A,

$$\limsup_{L \to \infty} \left| E_{\Lambda_L,(\beta J_e)}^{\bar{s}^{(L)}}(S_A) - E_{\Lambda_L,(\beta J_e)}^{\bar{t}^{(L)}}(S_A) \right| = 0. \quad (3.89)$$

We claim that this implies uniqueness for the (βJ_e) infinite volume Gibbs state, for if there were non-uniqueness there would be two sequences of b.c. measures $\bar{\rho}_L$ and $\bar{\rho}'_L$ on $\mathcal{S}_{\partial\Lambda_L}$ and some finite A such that

$$\lim_{L \to \infty} E_{\Lambda_L,(\beta J_e)}^{\bar{\rho}_L}(S_A) \neq \lim_{L \to \infty} E_{\Lambda_L,(\beta J_e)}^{\bar{\rho}'_L}(S_A), \quad (3.90)$$

which would allow one to extract $\bar{s}^{(L)}$ and $\bar{t}^{(L)}$ (from the supports of $\bar{\rho}_L$ and $\bar{\rho}'_L$) violating (3.89). This completes the proof. $\qquad\qquad\qquad\qquad\qquad\qquad \square$

Chapter 4

Low Temperature States of Disordered Systems

An Edwards-Anderson (EA) spin glass model [EA75] is a disordered Ising model on $\mathbb{Z}^d$ whose nearest neighbor couplings $J = (J_e : e \in \mathbb{E}^d)$ are i.i.d. random variables (on some $(\Omega, \mathcal{F}, \nu)$) with a common symmetric distribution μ (i.e., J_e and $-J_e$ are equidistributed). The most common examples are the Gaussian (where μ is a mean zero normal distribution) and the $\pm \breve{J}$ (where $\mu = \frac{1}{2}\delta_{\breve{J}} + \frac{1}{2}\delta_{-\breve{J}}$) models. We place no restrictions on μ beyond symmetry.

In the last chapter we defined two critical values β_c and $\bar{\beta}_c$ (see (3.48), Proposition 3.9 and (3.58)-(3.59)) which are nonrandom with $\beta_c \leq \bar{\beta}_c$. β_c and $\bar{\beta}_c$ are functions of d and μ. For $\beta' < \beta_c$, a.s. there is a unique infinite volume Gibbs state for all $\beta \leq \beta'$. On the supercritical side:

$$\beta' > \beta_c \implies \text{a.s. there is non-uniqueness for some } \beta < \beta', \quad (4.1)$$

$$\beta' > \bar{\beta}_c \implies \text{a.s. there is non-uniqueness for } \beta = \beta'. \quad (4.2)$$

We remark that there are other natural definitions of the critical β, but we will not discuss them here. For ferromagnetic models, $\beta_c = \bar{\beta}_c$, but no such result is known for more general couplings (including spin glasses). In particular, the possibility has not been ruled out that for some choice of d and μ, $\beta_c < \infty$ while $\bar{\beta}_c = \infty$. But even more basic issues about the values of β_c and $\bar{\beta}_c$ are mathematically unresolved. Before listing some of them, we give in the next proposition the known facts. Some of these are elementary, others follow from the results of Chapter 3 but have been obtained by other arguments in the literature, as noted in Chapter 3. As usual, we denote by $p_c(d)$, the critical value for standard independent Bernoulli bond percolation on the nearest neighbor graph $(\mathbb{Z}^d, \mathbb{E}^d)$.

Proposition 4.1 *For any d and μ, $\beta_c(d, \mu) > 0$. $\beta_c(d, \mu) = \infty$ for $d = 1$ and any μ as well as for $d \geq 2$ if $\mu(\mathbb{R} \setminus \{0\}) \leq p_c(d)$.*

Proof. These facts for $d \geq 2$ follow immediately from Theorems 3.10 and 3.11. The fact that $\beta_c(1, \mu) = \infty$ follows either from elementary one-dimensional arguments (which we leave for the reader, except for pointing out that a Gibbs

47

distribution on $\mathbb{Z}^1$ corresponds to an inhomogeneous two-state Markov chain) or else by noting that the $d \geq 2$ arguments of Theorems 3.10 and 3.11 also apply to $d = 1$, with $p_c(1) = 1$.   □

The general belief in the physics literature, based primarily on numerical studies, is that there is some critical dimension d_c (probably with $d_c \leq 3$ [FS90]) such that for $d \geq d_c$, the critical β is finite (at least for the standard examples of μ). It should be noted that in this literature, it is not usually clear which of the possible definitions of the critical value is under consideration. It is also believed that, like in the random field Ising model [AW90], the critical β is infinite for $d = 2$. Thus we have the following open problems.

Research Problem 4.2 *Determine (either for all μ or at least for some μ with $\mu(\mathbb{R} \setminus \{0\}) > p_c(2) = 1/2$) whether $\beta_c(2, \mu) = \infty$, or at least whether $\bar{\beta}_c(2, \mu) = \infty$.*

Research Problem 4.3 *Determine whether for some d and some μ, $\bar{\beta}_c(d, \mu) < \infty$, or at least whether $\beta_c(d, \mu) < \infty$.*

Although essentially nothing is known rigorously about β_c or $\bar{\beta}_c$ beyond what can be extracted from Theorems 3.10 and 3.11, there are some results concerning FK percolation that are related to the above two research problems. Note that the presence of FK percolation is required for spin flip symmetry breaking; i.e., its absence (for arbitrary b.c.'s) implies that every infinite volume Gibbs state is invariant under the mapping $s \to -s$. One result [DG96] concerns the $\pm \check{J}$ spin glass with $d = 2$, where $\beta_c \geq \beta_c^F$ (by Theorem 3.11) but there is no proof that $\beta_c > \beta_c^F$. The result is that for some $\epsilon > 0$ and any $\beta < \beta_c^F + \epsilon$, a.s. there is absence of FK percolation. In the other direction, it is known [GKN92] (see also [N94]) that for $d > 2$, $\mu(\{0\})$ small and β large, a.s. FK percolation is present (with stronger results available for the $\pm \check{J}$ spin glass [DG96]).

For a given ω and hence a given coupling configuration $J = J(\omega)$, we may consider the set $\mathcal{G} = \mathcal{G}(J(\omega), \beta)$ of all infinite volume Gibbs states for $J(\omega)$ at inverse temperature β. In analyzing $\mathcal{G}$ it is natural to consider the set of extremal (or pure) Gibbs states,

$$\text{ex } \mathcal{G} = \mathcal{G} \setminus \{\alpha P_1 + (1 - \alpha)P_2 : \alpha \in (0, 1); \ P_1, P_2 \in \mathcal{G}; \ P_1 \neq P_2\}. \qquad (4.3)$$

We define $N = N(J(\omega), \beta)$ to be the cardinality, $|\text{ex } \mathcal{G}|$, of ex $\mathcal{G}$. N can (a priori) take any of the values $1, 2, \ldots$, or ∞ (we do not distinguish here between countably and uncountably infinite). As part of the proof of Proposition 3.9, it was shown that $\{\omega : N = 1\}$ is measurable (i.e., belongs to $\mathcal{F}$ and in fact to the σ-field generated by J). The next proposition extends that fact.

Proposition 4.4 *$N(J(\omega), \beta)$ is measurable and a.s. equals a constant, $\mathcal{N}(d, \mu, \beta)$.*

Proof. We will use the fact from the general theory of Gibbs distributions (see Corollary 7.28 of [Ge88]) that $N \geq k$ if and only if $\mathcal{G}$ contains at least k linearly independent measures $P_1, \ldots, P_k$. But this is so if and only if for some finite $\Lambda \subset \mathbb{Z}^d$, there is linear independence of the ($2^{|\Lambda|}$-dimensional) vectors $\vec{p}_\Lambda(P_1), \ldots, \vec{p}_\Lambda(P_k)$, defined as

$$\vec{p}_\Lambda(P) = (P(s = t \text{ on } \Lambda) : t \in \mathcal{S}_\Lambda). \tag{4.4}$$

Thus $N \geq k$ if and only if for some finite Λ, the closed convex set,

$$G_\Lambda \equiv \{\vec{p}_\Lambda(P) : P \in \mathcal{G}\}, \tag{4.5}$$

is at least k-dimensional.

For L sufficiently large (so that Λ is contained in the cube Λ_L), let us consider the finite volume approximation to G_Λ, obtained by replacing $\mathcal{G}$ in (4.5) by the finite volume Gibbs measures $P^{\bar\rho}_{\Lambda_L,\beta}$:

$$G^L_\Lambda \equiv \left\{\vec{p}_\Lambda(P^{\bar\rho}_{\Lambda_L,\beta}) : \bar\rho \text{ is a probability measure on } \mathcal{S}_{\partial\Lambda_L}\right\}. \tag{4.6}$$

Then $G^{L''}_\Lambda \subseteq G^{L'}_\Lambda$ for $L'' > L'$ and $G_\Lambda = \lim_{L\to\infty} G^L_\Lambda$. Let $\mathcal{R}_{\Lambda,k}$ denote the countable set of linearly independent k-tuples $(\vec{p}_1, \ldots \vec{p}_k)$ of vectors in $\mathbb{R}^{2^{|\Lambda|}}$ with rational coordinates. Then G_Λ is at least k-dimensional if and only if there exists $(\vec{p}_1, \ldots \vec{p}_k) \in \mathcal{R}_{\Lambda,k}$ such that for each j, $\vec{p}_j \in G^L_\Lambda$ for all (large) L.

Next we note that since $P^{\bar\rho}_{\Lambda_L,\beta}$ is just the convex combination (corresponding to $\bar\rho$) of the fixed b.c. $P^{\bar s}_{\Lambda_L,\beta}$'s,

$$G^L_\Lambda = \text{convex hull } \left(\vec{p}_\Lambda(P^{\bar s}_{\Lambda_L,\beta}) : \bar s \in \mathcal{S}_{\partial\Lambda_L}\right). \tag{4.7}$$

From the definition of $P^{\bar s}_{\Lambda_L,\beta}$ we clearly have that the countably many points $\vec{p}_\Lambda(P^{\bar s}_{\Lambda_L,\beta})$, as Λ, L and $\bar s$ vary, are all measurable functions of J. Thus

$$\begin{aligned}
\{N \geq k\} = \; & \big\{\exists\Lambda, \exists(\vec{p}_1, \ldots \vec{p}_k) \in \mathcal{R}_{\Lambda,k}, \\
& \forall j, \; \forall \text{ large } L, \\
& \vec{p}_j \in \text{convex hull } \big(\vec{p}_\Lambda(P^{\bar s}_{\Lambda_L,\beta}) : \bar s \in \mathcal{S}_{\partial\Lambda_L}\big)\big\}
\end{aligned} \tag{4.8}$$

is measurable.

Finally, to complete the proof, note that the joint distribution $\bar\nu$ of $J = (J_e : e \in \mathbb{E}^d)$ is an i.i.d. product measure which is invariant and ergodic with respect to translations on $\mathbb{Z}^d$. Since N is a measurable function of J which is clearly translation invariant, it follows that N is a.s. constant. $\qquad\square$

For a given J and β, $N \geq 2$ if and only if there is non-uniqueness of infinite volume Gibbs distributions. Thus the issue of whether $\bar\beta_c(d, \mu) < \infty$ is equivalent to whether $\mathcal{N}(d, \mu, \beta) \geq 2$ for all large β. Although it seems to be generally believed that (at least for some μ's), $\mathcal{N}(d, \mu, \beta) \geq 2$ for $d \geq$ some d_c and large β, there is a lively controversy in the physics literature on the nature of the pure Gibbs states and, in particular, on the value of $\mathcal{N}$ (when $\mathcal{N} \geq 2$).

One side of the controversy is fairly easy to describe. This side predicts, on the basis of nonrigorous scaling arguments [M84], [FH86], [BrM87], [FH88], that

$\mathcal{N} = 2$ and that the two pure states, P' and P'', are global flips of each other, i.e., that the mapping $s \to -s$ on $\mathcal{S}$ transforms P' into P'' and vice versa. (See [BoF86] for different arguments that predict $\mathcal{N} = 2$ only for $d = 3$ and [En90] for a critique of the scaling argument prediction that $\mathcal{N} = 2$ for all d.) This would be analogous to the situation for the homogeneous $d = 2$ Ising ferromagnet [A80], [Hi81]. The local magnetizations $\langle S_x \rangle'$ (where $\langle \cdot \rangle'$ denotes expectation w.r.t. P') and $\langle S_x \rangle'' = -\langle S_x \rangle'$ would depend on both x and ω and would be nonvanishing (at least for some x's with a positive density in $\mathbb{Z}^d$). The mean magnetization should vanish; i.e., w.r.t. P',

$$\lim_{L\to\infty} |\Lambda_L|^{-1} \sum_{y\in\Lambda_L} S_y = \lim_{L\to\infty} |\Lambda_L|^{-1} \sum_{y\in\Lambda_L} \langle S_y \rangle' = 0. \qquad (4.9)$$

However, the EA order parameter should be a strictly positive constant, $r_{EA} = r_{EA}(d, \mu, \beta)$, given by

$$r_{EA} = \lim_{L\to\infty} |\Lambda_L|^{-1} \sum_{y\in\Lambda_L} [\langle S_y \rangle']^2 = E\left([\langle S_x \rangle']^2\right), \qquad (4.10)$$

where E denotes expectation with respect to ν (on the probability space of J).

One feature of the other side of the controversy is also easy to describe — i.e., the prediction that $\mathcal{N} = \infty$. Other features are more difficult and most of this chapter will be devoted to presenting and analyzing different interpretations of how $\mathcal{N} = \infty$ could or should manifest itself. Before we begin that, we list an obvious open problem.

Research Problem 4.5 *Determine whether $\mathcal{N}(d, \mu, \beta) = \infty$, or at least whether $\mathcal{N}(d, \mu, \beta) > 2$, for some d, μ and β.*

The prediction that $\mathcal{N} = \infty$ is based firstly on the notion that a short-ranged spin glass like the EA model should behave qualitatively like an "infinite-ranged" or "mean-field" model and based secondly on the nonrigorous analysis by Parisi and others [P79], [P83], [HoJY83], [MPSTV84] of the Sherrington-Kirkpatrick (SK) model [SK75], the standard mean-field spin glass. This line of reasoning leads not only to the prediction that $\mathcal{N} = \infty$ (for appropriate d, μ and β) but also to many other predictions concerning the Gibbs states for large finite volume and for infinite volume.

These predictions, taken as a whole, constitute what we call an SK picture of the EA model. Unfortunately the predictions (other than that $\mathcal{N} = \infty$) have not been formulated in the literature with much precision so that before attacking the problem of whether the SK picture (of the EA model) is valid, it is first necessary to consider the question, "what is the SK picture?". One of our main objectives here, following [NS92], [NS96b], [NS96c], [NS96d], [NS97], is to answer that question. To begin, we give a very brief introduction to the SK model itself and to Parisi's analysis of it. For a comprehensive discussion, the reader is referred to [MPV87].

Unlike all the Ising models we have discussed so far, the SK model has no a priori connection with the lattice $\mathbb{Z}^d$ and so it is conventional to replace subsets of $\mathbb{Z}^d$ by subsets of $\mathbb{N}$. The SK model of size n at inverse temperature β has the Gibbs measure on $\mathcal{S}_n = \{-1, +1\}^n$,

$$P_{n,\beta}(\{s\}) = Z_{n,\beta}^{-1} \exp\left(\beta \sum_{i=1}^{n-1} \sum_{j=i+1}^{n} J_{ij}^n s_i s_j\right), \qquad (4.11)$$

with the n-dependent couplings given by

$$J_{ij}^n = n^{-1/2} K_{ij} \quad \text{for } 1 \le i < j < \infty, \qquad (4.12)$$

where the K_{ij}'s are i.i.d. random variables (on some $(\Omega, \mathcal{F}, \nu)$) with common symmetric distribution μ.

This Gibbs measure may be regarded, in an obvious sense, as having a free b.c. In another sense, since there is no natural notion of a boundary at all in the SK model, it may be regarded as more analogous to taking periodic b.c.'s in nearest neighbor models, since those are generally thought of as minimizing the effects of the boundary. There seems to be no obvious analogue in the SK model to nearest neighbor model b.c.'s such as fixing $\bar{s}$ on $\partial\Lambda$.

The ferromagnetic infinite-ranged or mean-field model, known as the Curie-Weiss (CW) model, has a Gibbs distribution also of the form of (4.11) but where (4.12) is replaced by $J_{ij}^n = n^{-1}$. In both SK and CW models, the scaling factors, $n^{-1/2}$ and n^{-1}, are chosen so that the mean energy per site,

$$n^{-1} \left\langle -\sum_{i=1}^{n-1} \sum_{j=i}^{n} J_{ij}^n s_i s_j \right\rangle_{n,\beta}, \qquad (4.13)$$

is bounded away from 0 and ∞; the difference in scaling exponents is due to the presence of sign cancellations in the SK model and their absence in the CW model. The CW model has a simple and easily derived exact solution — e.g., for the (infinite volume limit) free energy per site,

$$f(\beta) = \lim_{n \to \infty} -\frac{1}{n\beta} \ln Z_{n,\beta}. \qquad (4.14)$$

This is not the case for the SK model.

In the CW model (as in the homogeneous (nearest-neighbor) ferromagnet on $\mathbb{Z}^d$), a natural object to study is the magnetization per site, i.e., the random variable (on $\mathcal{S}_n$ with probability measure $P_{n,\beta}$), $n^{-1} \sum_{i=1}^{n} s_i$. The derivation of the exact formula for the free energy also yields information on the $n \to \infty$ limit of the distribution of this random variable. There is a critical value β_c^{CW} such that for $\beta \le \beta_c^{CW}$, the limiting distribution is simply δ_0, the point measure at the origin, while for $\beta > \beta_c^{CW}$, the limit is $\frac{1}{2}\delta_{M^*} + \frac{1}{2}\delta_{-M^*}$, where $M^* = M^*(\beta)$, the spontaneous magnetization, is strictly positive for $\beta > \beta_c^{CW}$. In the SK model, the magnetization per site is not a fruitful choice of variable since, due to sign cancellations, it should have a trivial limiting distribution, δ_0, for all β. Indeed, Parisi's analysis of the free energy for the SK model led him to study the distribution of a different random variable, the replica overlap.

The replica overlap is the random variable (on $\mathcal{S}_n \times \mathcal{S}_n = \{(s^1, s^2)\}$ with probability measure $P_{n,\beta} \times P_{n,\beta}$)

$$R_n = \frac{1}{n} \sum_{i=1}^{n} s_i^1 s_i^2. \tag{4.15}$$

The formal density of its distribution (on the interval $[-1, 1]$) is denoted $\mathcal{P}_n(r)$. For β below a critical value β_c^{SK}, the $n \to \infty$ limit of $\mathcal{P}_n(r)$ is $\delta(r)$, the trivial point density at 0. But above β_c^{SK}, Parisi found highly nontrivial behavior: namely, that as $n \to \infty$, $\mathcal{P}_n$ approximates a sum of many delta functions, at locations and with weights which depend on ω, with the weights *not* tending to zero and with the dependence of the locations and weights on ω also *not* tending to zero as $n \to \infty$.

The usual explanation given for this behavior (see, e.g., [MPV87]) is that, as $n \to \infty$, the Gibbs measure $P_{n,\beta}$ has a decomposition into many pure states $P^\alpha = P_\beta^\alpha$,

$$P_{n,\beta} \approx \sum_\alpha W^\alpha P^\alpha, \tag{4.16}$$

with weights W^α depending on ω, with neither they nor their dependence on ω tending to zero. [Warning: The reader should not be alarmed if the meaning of a pure state in the SK context is not precisely clear; it is certainly not clear to this writer.]

If the replicas s^1 and s^2 were chosen from $P^\alpha \times P^\gamma$ rather than from $P_{n,\beta} \times P_{n,\beta}$, then, as $n \to \infty$, the overlap would have a point density

$$\delta(r - r_{\alpha\gamma}), \qquad r_{\alpha\gamma} \approx \frac{1}{n} \sum_{i=1}^{n} \langle S_i \rangle^\alpha \langle S_i \rangle^\gamma, \tag{4.17}$$

and so

$$\mathcal{P}_n(r) \approx \sum_\alpha \sum_\gamma W^\alpha W^\gamma \delta(r - r_{\alpha\gamma}). \tag{4.18}$$

Like the W^α's, the $r_{\alpha\gamma}$'s depend on ω with a dependence not tending to zero. This persistence in the dependence of various quantities on ω in the infinite volume limit, $n \to \infty$, is called non-self-averaging (NSA) and is one of the essential features of Parisi's analysis. (For rigorous approaches to NSA in the SK model, see [PS91], [T96] and [PST97].)

A second essential feature is that there are many α's (with non-negligible weights) appearing in (4.16) so that the two replicas have non-negligible probability of appearing in different (and "unrelated") pure states α and γ (with overlap value

$r_{\alpha\gamma}$). This is called replica symmetry breaking (RSB). By unrelated, we mean that γ is neither α nor the "negative" of α, i.e., that P^γ is not P^α and is also not the global spin flip of P^α. When $\gamma = \alpha$, $r_{\alpha\alpha}$ should, for all α, be the Edwards-Anderson order parameter r_{EA} (the analogue of (4.10), but for the SK model), while for $\gamma = -\alpha$, $r_{\alpha\gamma} = -r_{EA}$. The value of r_{EA} (unlike other $r_{\alpha\gamma}$'s) should *not* depend on ω.

A third essential feature, closely related to NSA, is that the discrete nature of the Parisi order parameter distribution as a countable sum of delta functions is only for fixed ω. As ω varies, the $r_{\alpha\gamma}$'s vary in such a way that, as $n \to \infty$, $\bar{P}_n(r)$, the average over the couplings (i.e., over the underlying probability measure ν on Ω) is continuous (except for two delta functions at $\pm r_{EA}$ since those locations do not depend on ω).

A fourth essential feature, closely related to, but going well beyond, the discreteness of the order parameter distribution, is ultrametricity of the $r_{\alpha\gamma}$'s (for fixed ω). Here one regards $d_{\alpha\gamma} \equiv r_{EA} - r_{\alpha\gamma}$ as defining a metric on the pure states. The ultrametric property of $d_{\alpha\gamma}$ is that among any three pure states α, γ and δ, the largest two of $d_{\alpha\gamma}$, $d_{\alpha\delta}$, $d_{\gamma\delta}$ are equal. This property occurs naturally, for example, in the context where the points in the ultrametric space correspond to valleys in a landscape and the distance corresponds to the height of the lowest pass that must be traversed to travel from one valley to another.

In formulating a precise interpretation of the SK picture of the EA model, it is clear that a primary role will be played by an EA model analogue of the approximate pure state decomposition (4.16). In our EA analogues of (4.16), we will replace $P_{n,\beta}$ on the LHS by $P_L = P_{\Lambda_L,\beta}^{per}$, the ($\omega$-dependent) Gibbs measure for the EA model on the cube Λ_L with periodic b.c.'s. (For many purposes, free b.c.'s or, indeed, any b.c. not depending on ω could equally well be chosen; it should be noted however that there is some possibility of different behavior between periodic and, say, free b.c.'s [En90].) The EA model has a great advantage over the SK model in that there is already a well defined meaning of pure states as extremal infinite volume Gibbs states. Thus in the EA model, the role of pure states will be played by ... the pure states. The real issue for EA analogues of (4.16) is interpreting the approximate equality.

The most straightforward interpretation of the SK picture, which we call the standard SK picture, is also the one that most closely matches the presentations in the physics literature. In this picture, (4.16) is replaced by the identity

$$P_{\mathcal{J}} = \sum_\alpha W_{\mathcal{J}}^\alpha P_{\mathcal{J}}^\alpha, \qquad (4.19)$$

where $\mathcal{J} = (\mathcal{J}_e : e \in \mathbb{E}^d)$ represents a particular configuration of the random couplings $J = (J_e : e \in \mathbb{E}^d)$, $P_{\mathcal{J}}$ is an infinite volume Gibbs state for $\mathcal{J}$ (and some fixed β) obtained in some natural way from the finite volume periodic b.c. Gibbs states $P_L = P_{\mathcal{J},L}$ by letting $L \to \infty$, and the $P_{\mathcal{J}}^\alpha$'s are pure (i.e., extremal infinite volume) Gibbs states for that same $\mathcal{J}$. The identity (4.19) is to be valid for $\bar{\nu}$-a.e.

$\mathcal{J}$, where $\bar{\nu}$ is the joint distribution of the J_e's. The replica overlap in this picture is the random variable (on $\mathcal{S} \times \mathcal{S}$ with probability measure $P_{\mathcal{J}} \times P_{\mathcal{J}}$),

$$R_{\mathcal{J}} = \lim_{L \to \infty} |\Lambda_L|^{-1} \sum_{x \in \Lambda_L} s_x^1 s_x^2, \qquad (4.20)$$

and (the formal density of) its distribution is the Parisi overlap distribution $\mathcal{P}_{\mathcal{J}}$. The limit over cubes in (4.20) is to exist, a.s. with respect to $P_{\mathcal{J}} \times P_{\mathcal{J}}$, for $\bar{\nu}$-a.e. $\mathcal{J}$ and the overlap distribution $\mathcal{P}_{\mathcal{J}}$ is to depend measurably on $\mathcal{J}$.

If all this were so, then it would follow from (4.19) and the tail σ-field triviality of pure Gibbs states (see Chapter 7 of [Ge88]), that

$$\mathcal{P}_{\mathcal{J}}(r) = \sum_{\alpha} \sum_{\gamma} W_{\mathcal{J}}^{\alpha} W_{\mathcal{J}}^{\gamma} \delta(r - r_{\alpha\gamma}^{\mathcal{J}}), \qquad (4.21)$$

with

$$r_{\alpha\gamma}^{\mathcal{J}} = \lim_{L \to \infty} |\Lambda_L|^{-1} \sum_{x \in \Lambda_L} \langle S_x \rangle_{\mathcal{J}}^{\alpha} \langle S_x \rangle_{\mathcal{J}}^{\gamma}. \qquad (4.22)$$

The standard SK picture thus predicts that the overlap distribution $\mathcal{P}_{\mathcal{J}}$, obtained from $P_{\mathcal{J}}$ in this way, has the following four essential features. (1) NSA $-$ $\mathcal{P}_{\mathcal{J}}$ does depend on $\mathcal{J}$; (2) nontrivial discreteness $-$ $\mathcal{P}_{\mathcal{J}}$ is a sum of (countably) infinitely many delta functions; (3) continuity of its $\mathcal{J}$-average $-$ $\bar{\mathcal{P}} \equiv \int \mathcal{P}_{\mathcal{J}} d\bar{\nu}(\mathcal{J})$ is continuous (except for two δ-functions at $\pm r_{EA}$, whose weights add up to less than one); (4) ultrametricity of the $r_{\alpha\gamma}^{\mathcal{J}}$'s. (We remark that although discreteness of $\mathcal{P}_{\mathcal{J}}$ is essential to this SK picture, it has been pointed out to us by A. van Enter that, in principle, this could be the case without discreteness of the pure state decomposition (4.19); indeed, such a situation occurs in some deterministic models considered in [EnHM92].)

Now that we have formulated the standard SK picture, we can ask whether this picture of the EA model can be valid (for some dimensions and some temperatures). This question has two parts. First, does there exist some natural construction which begins with the finite volume states, $P_{\mathcal{J},L}$, takes $L \to \infty$, and ends with an infinite volume state, $P_{\mathcal{J}}$, and its accompanying overlap distribution $\mathcal{P}_{\mathcal{J}}$? Second, can such a $\mathcal{P}_{\mathcal{J}}$ exhibit all the essential features of the SK picture? The answers to these two parts, given in [NS96b] are, respectively, yes and no, as we now explain. We will not formalize the answer to the first part as a proposition or theorem because it will be implicitly included (in Theorem 4.8 and Remark 4.9) as a part of the more comprehensive "metastate" approach given later in this chapter.

We begin our answer to the first part of the question by noting that we cannot simply fix $\mathcal{J}$ and take an ordinary limit of the finite cube, periodic b.c. state $P_{\mathcal{J},L}$, as $L \to \infty$. Unlike, say, the $d = 2$ homogeneous Ising ferromagnet, where such a limit exists (and equals $\frac{1}{2}P^+ + \frac{1}{2}P^-$) by spin flip symmetry considerations (and the fact that P^+ and P^- are the only pure states [A80], [Hi81]), there is no guarantee for a spin glass that there is a well defined limit. By compactness arguments, one can easily obtain, for each $\mathcal{J}$, convergence along subsequences of L's. But these subsequences may be $\mathcal{J}$-dependent and there seems to be no natural way to patch together the limits for different $\mathcal{J}$'s to yield $P_{\mathcal{J}}$. This is more than a technical problem. As first discussed in [NS92] and as we will explain below, when there are many competing pure states (as in an SK picture), there should be chaotic size dependence (CSD) – i.e., the existence, for typical configurations of $\mathcal{J}$, of different limits along different ($\mathcal{J}$-dependent) subsequences.

In spite of the "problem" of CSD, a limit can be taken, not by fixing $\mathcal{J}$, but by considering the joint distribution $\bar{\nu}(\mathcal{J}) \times P_{\mathcal{J},L}((s_x : x \in \Lambda_L))$, of the J_e's and the S_x's. (We note that such joint distribution limits were considered, implicitly or explicitly, in [Le77], [Co89], [GKN92] and [S95].) That is, by choosing a subsequence of L's (not depending on $\mathcal{J}$), one has convergence of all the finite dimensional distributions of the J_e's and S_x's to those of a probability measure on $\mathbb{R}^{\mathbb{E}^d} \times \mathcal{S}$, whose marginal distribution of $\mathcal{J}$ is $\bar{\nu}$ and whose conditional distribution of s, given $\mathcal{J}$, is some $P_{\mathcal{J}}$. All this follows from standard compactness arguments with $P_{\mathcal{J}}$ a probability measure on $\mathcal{S}$ defined for $\bar{\nu}$-a.e. $\mathcal{J}$ and depending measurably on $\mathcal{J}$. What doesn't follow from general compactness arguments is that (for $\bar{\nu}$-a.e. $\mathcal{J}$), $P_{\mathcal{J}}$ is an infinite volume Gibbs state for $\mathcal{J}$. This however can be shown either directly, or as a corollary of a more comprehensive result from [AW90] that is also presented below (see Theorem 4.8 and Remark 4.9).

This construction has certain translation invariance properties that are important, both technically and conceptually. Because of the periodic b.c.'s on the cube Λ_L, the couplings and spins are really defined on a (discrete) torus of size L, with a joint distribution invariant under torus translations. This implies that any (subsequence) limit joint distribution on $\mathbb{R}^{\mathbb{E}^d} \times \mathcal{S}$ is invariant under translations of $\mathbb{Z}^d$, which in turn implies that $P_{\mathcal{J}}$ is translation covariant; i.e., under the translation of $\mathcal{J}$ to $\mathcal{J}^a$, where $\mathcal{J}^a_{\{x,y\}} = \mathcal{J}_{\{x+a,y+a\}}$, $P_{\mathcal{J}}$ transforms so that

$$P_{\mathcal{J}^a}(S_{x_1} = s_1, \ldots, S_{x_m} = s_m) = P_{\mathcal{J}}(S_{x_1-a} = s_1, \ldots S_{x_m-a} = s_m). \qquad (4.23)$$

The conceptual significance of translation covariance is that the mapping from $\mathcal{J}$ to $P_{\mathcal{J}}$, being a natural one, should not (and in this construction does not) depend on the choice of an origin. The technical significance is that it implies that the joint measure for $\mathcal{J}$ and the two replicas s^1 and s^2, $\bar{\nu}(\mathcal{J})P_{\mathcal{J}}(s^1)P_{\mathcal{J}}(s^2)$, on $\mathbb{R}^{\mathbb{E}^d} \times \mathcal{S} \times \mathcal{S}$ is translation invariant (under $(\mathcal{J}, S^1, S^2) \to (\mathcal{J}^a, S^{1a}, S^{2a})$) which implies, by the ergodic theorem, that $|\Lambda_L|^{-1} \sum_{x \in \Lambda_L} s^1_x s^2_x$ has an a.s. limit R (with respect to $\bar{\nu}(\mathcal{J})P_{\mathcal{J}}(s^1)P_{\mathcal{J}}(s^2)$) and thus the $R_{\mathcal{J}}$ of (4.20) exists a.s. (with respect to $P_{\mathcal{J}} \times P_{\mathcal{J}}$) for $\bar{\nu}$-a.e. $\mathcal{J}$, as desired. $\mathcal{P}_{\mathcal{J}}$, the distribution of $R_{\mathcal{J}}$, then exists for $\bar{\nu}$-a.e. $\mathcal{J}$ and depends measurably on $\mathcal{J}$. Indeed $\mathcal{P}_{\mathcal{J}}$ is simply the conditional distribution of the random variable R, given $\mathcal{J}$.

We have answered the first part of our question on the validity of the standard SK picture by showing that, yes, there does exist a natural $P_{\mathcal{J}}$ and $\mathcal{P}_{\mathcal{J}}$, that are related as required by that picture and that depend on $\mathcal{J}$ measurably. To begin our answer to the second part of the question, we see what the translation covariance of $P_{\mathcal{J}}$ implies about $\mathcal{P}_{\mathcal{J}}$. By translation covariance, $R_{\mathcal{J}^a}$ is equidistributed with the random variable (on $\mathcal{S} \times \mathcal{S}$ with measure $P_{\mathcal{J}} \times P_{\mathcal{J}}$)

$$R_{\mathcal{J}}^{-a} \equiv \lim_{L \to \infty} |\Lambda_L|^{-1} \sum_{x \in \Lambda_L} s_{x-a}^1 s_{x-a}^2 = R_{\mathcal{J}}. \qquad (4.24)$$

Thus $\mathcal{P}_{\mathcal{J}^a} = \mathcal{P}_{\mathcal{J}}$ for $\bar{\nu}$-a.e. $\mathcal{J}$ and all $a \in Z^d$; i.e., $\mathcal{P}_{\mathcal{J}}$ is translation invariant. As in the case of the translation covariance of $P_{\mathcal{J}}$, the translation invariance of $\mathcal{P}_{\mathcal{J}}$ has the conceptual significance that a natural object like the Parisi order parameter distribution should not (and in this construction does not) depend on the choice of an origin. But it also has an important technical significance which, in the next proposition, explains why the answer to the second part of our question on the validity of the standard SK picture is, no, such a $\mathcal{P}_{\mathcal{J}}$ cannot exhibit all the essential features of the SK picture.

Proposition 4.6 ([NS96b]) *If $\mathcal{P}_{\mathcal{J}}$ is translation invariant, then it is self-averaged; i.e., it equals a fixed probability measure $\mathcal{P}$ on $[-1,1]$ for $\bar{\nu}$-a.e. $\mathcal{J}$. Thus it does not exhibit (1) non-self-averaging and consequently also does not exhibit at least one of (2) nontrivial discreteness or (3) continuity of its $\mathcal{J}$-average.*

Proof. Consider for any k, the moment $\int_{-1}^{1} q^k \mathcal{P}_{\mathcal{J}}(q) dq$. This is a measurable function of $\mathcal{J}$ defined for $\bar{\nu}$-a.e. $\mathcal{J}$ which is invariant under the translation $\mathcal{J} \to \mathcal{J}^a$ for every $a \in \mathbb{Z}^d$. By the spatial translation invariance and ergodicity of $\bar{\nu}$, this implies that this function is $\bar{\nu}$-a.s. a constant. These moments, for all k, determine $\mathcal{P}_{\mathcal{J}}$; thus $\mathcal{P}_{\mathcal{J}}$ itself is $\bar{\nu}$-a.s. a constant $\mathcal{P}$. Since $\mathcal{P} = \int \mathcal{P}_{\mathcal{J}} d\bar{\nu}(\mathcal{J})$, the rest of the proposition follows. $\qquad \square$

Technically, the feature of the standard SK picture which led to its demise, was the translation covariance of the infinite volume state $P_{\mathcal{J}}$. In pursuing other interpretations of the SK picture, and in analyzing disordered systems more generally, we will not give up translation covariance, but we will give up the idea that a disordered system, in the infinite volume limit, should necessarily be described by a single $P_{\mathcal{J}}$, i.e., by a function from coupling configurations to *single* infinite volume Gibbs states. Indeed, rather than finessing CSD (by constructing our $P_{\mathcal{J}}$), we will try to understand (or at least describe) CSD by analyzing the way in which $P_{\mathcal{J},L}$ samples from its various possible limits as $L \to \infty$. A major contribution of [NS96c] was the proposal that this sampling is naturally understood in terms of a "metastate", a probability measure $\kappa_{\mathcal{J}}$ on the infinite volume Gibbs states for the given $\mathcal{J}$. (We will give a more precise definition below.)

This proposal of [NS96c] was based on an analogy with chaotic deterministic dynamical systems, where the chaotic motion along a deterministic orbit is analyzed in terms of some appropriately selected probability measure, invariant under the dynamics. Time along the orbit is replaced, in our context, by L and the state space (or configuration space or phase space) of the dynamical system is replaced by the space of Gibbs states (for a fixed $\mathcal{J}$). We will not pursue here the issue of what, in disordered systems, replaces invariance of the probability measure under the dynamics. Rather we will explain how the same metastate can be constructed by two different approaches, one based on the randomness of the $\mathcal{J}$'s and the other based on CSD for a fixed $\mathcal{J}$.

The approach based on $\mathcal{J}$-randomness is due to Aizenman and Wehr [AW90]. This approach is analogous to the construction of $P_{\mathcal{J}}$ described above, except that instead of considering the random pair (J, S), distributed for finite L by $\bar{\nu}(\mathcal{J}) \times P_{\mathcal{J},L}$, and then taking the limit (along a subsequence) of finite dimensional distributions, one considers the random pair $(J, P_{J,L})$, defined on the underlying probability space $(\Omega, \mathcal{F}, \nu)$ of J, and takes the limit of its finite dimensional distributions. The finite dimensional distributions can be defined in a number of equivalent ways. We will consider (for $P_{J,L}$) the random (because of J) probabilities of cylinder sets. I.e., for each finite $\Lambda \subset \mathbb{Z}^d$ and $s \in \mathcal{S}_\Lambda = \{-1, +1\}^\Lambda$ (we denote by $\mathcal{A}$ the set of all such pairs (Λ, s)) and for L sufficiently large so that $\Lambda \subseteq \Lambda_L$, we define the random variable (on $(\Omega, \mathcal{F}, \nu)$)

$$Q_{(\Lambda,s)}^{(L)} = P_{J,L}\left(\{s' \in \mathcal{S}_{\Lambda_L} : s'_x = s_x \ \forall x \in \Lambda\}\right). \tag{4.25}$$

Let $\kappa^\dagger$ denote a probability measure on $\Omega^\dagger = \mathbb{R}^{\mathbb{E}^d} \times \mathbb{R}^{\mathcal{A}}$, with the product Borel σ-field $\mathcal{F}^\dagger$. We say that $(J, P_{J,L}) \to \kappa^\dagger$ as $L \to \infty$, if each of the finite dimensional distributions of $\left(J_e, Q_{(\Lambda,s)}^{(L)} : e \in \mathbb{E}^d, (\Lambda, s) \in \mathcal{A}\right)$ converges as $L \to \infty$ to the corresponding finite dimensional marginal of $\kappa^\dagger$.

Research Problem 4.7 *Prove that $(J, P_{J,L}) \to \kappa^\dagger$ for some probability measure $\kappa^\dagger$ on $(\Omega^\dagger, \mathcal{F}^\dagger)$.*

Although convergence of $(J, P_{J,L})$ has not been proved, it can be shown [AW90] (see also Lemmas B.2, B.3 and B.4 of Appendix B below) that there is sequential compactness and that every subsequence limit $\kappa^\dagger$ has a conditional distribution $\kappa_{\mathcal{J}}$ (of $q \in \Omega^1 \equiv \mathbb{R}^{\mathcal{A}}$, given $\mathcal{J} \in \Omega^0 \equiv \mathbb{R}^{\mathbb{E}^d}$) that, for $\bar{\nu}$-a.e. $\mathcal{J}$, is supported on infinite volume Gibbs distributions for that $\mathcal{J}$. Thus we have the following result.

Theorem 4.8 ([AW90]) *There exists a subsequence L_n of the L's such that $(J, P_{J,L_n}) \to \kappa^\dagger$. Here $\kappa^\dagger$ is a probability measure on $(\Omega^\dagger, \mathcal{F}^\dagger)$ whose marginal distribution $\kappa_{\mathcal{J}}$ (of q, given $\mathcal{J}$) satisfies:*

$$\textit{for } \bar{\nu}\textit{-a.e. } \mathcal{J}, \quad \kappa_{\mathcal{J}}\left(\{q : q \textit{ is an infinite volume Gibbs state for } \mathcal{J}\}\right) = 1. \tag{4.26}$$

Proof. This follows directly from Lemmas B.2, B.3 and B.4 of Appendix B. $\square$

Remark 4.9 There is more than an analogy relating the construction of $\kappa_{\mathcal{J}}$ given in the last proposition to the construction of the Gibbs distribution $P_{\mathcal{J}}$ given earlier. By restricting the function f appearing in (B.8) of Appendix B to be linear in q, we see that, as $n \to \infty$,

$$E\big(g(J)\langle S_A \rangle_{J,L_n}\big) \to \int_{\Omega^0} g(\mathcal{J}) \left[\int_{\Omega^1} \langle S_A \rangle_q d\kappa_{\mathcal{J}}(q) \right] d\bar{\nu}(\mathcal{J}), \qquad (4.27)$$

where g is a (continuous, bounded) function of finitely many couplings, A is a finite subset of $\mathbb{Z}^d$, $S_A = \prod_{x \in A} S_x$, $\langle \cdot \rangle_{J,L_n}$ denotes expectation with respect to P_{J,L_n}, and $\langle \cdot \rangle_q$ denotes the expectation with respect to q (for q a probability measure on $\mathcal{S}$). On the other hand, the construction of $P_{\mathcal{J}}$ yields

$$E\big(g(J)\langle S_A \rangle_{J,L_n}\big) \to \int_{\Omega^0} g(\mathcal{J})\langle S_A \rangle_{P_{\mathcal{J}}} d\bar{\nu}(\mathcal{J}). \qquad (4.28)$$

Since these last two equations are valid for general g and A, we conclude (see [AW90]) that the state $P_{\mathcal{J}}$ is the mean of the metastate $\kappa_{\mathcal{J}}$, i.e.,

$$P_{\mathcal{J}} = \int_{\Omega^1} q \, d\kappa_{\mathcal{J}}(q). \qquad (4.29)$$

(4.26) then implies that $P_{\mathcal{J}}$ is indeed a Gibbs distribution.

The second approach to constructing a metastate takes a fixed $\mathcal{J}$ and replaces $\mathcal{J}$-randomness, roughly speaking, by regarding L as random, i.e., by considering the empirical distribution of $P_{J,L}$ as L varies. We define the empirical distribution along a given subsequence L_n of the L's as follows. First, we make the convention that $Q^{(L)}_{(\Lambda,s)}$, defined by (4.25) for $\Lambda \subseteq \Lambda_L$, is zero when $\Lambda \not\subseteq \Lambda_L$, and then we define $\vec{Q}^{(L)}$ as the Ω^1-valued random variable (on the underlying probability space $(\Omega, \mathcal{F}, \nu)$ of J),

$$\vec{Q}^{(L)} = \left(Q^{(L)}_{(\Lambda,s)} : (\Lambda, s) \in \mathcal{A} \right). \qquad (4.30)$$

The empirical distribution, along the subsequence (L_n), is the random discrete measure on Ω^1,

$$\kappa^K_J \left((L_n) \right) = \frac{1}{K} \sum_{k=1}^{K} \delta_{\vec{Q}^{(L_k)}}. \qquad (4.31)$$

As in the first approach to metastates, the empirical distribution approach has an open problem about convergence along with a partial solution based on taking subsequences. For probability measures κ'_m and κ' on Ω^1, we say $\kappa'_m \to \kappa'$ if each of the finite dimensional marginals of κ'_m converges as $m \to \infty$ to the corresponding marginal of κ'.

Research Problem 4.10 *Determine whether for ν-a.e. ω, $\kappa^K_{J(\omega)}((L_n \equiv n)) \to \kappa'_\omega$, as $K \to \infty$, for some probability measure κ'_ω on Ω^1.*

The partial solution to this problem is based on Theorem 4.8. If (L_n) is a subsequence such that $(J, P_{J,L_n}) \to \kappa^\dagger$, it can be shown (see Proposition B.6 of Appendix B) that by taking a subsequence n_k of the n's and a subsequence K_m of the K's, that for ν-a.e. ω, $\kappa_{J(\omega)}^{K_m}((L_{n_k}))$ converges as $m \to \infty$ to $\kappa_{J(\omega)}$, where κ_J is the conditional distribution of $\kappa^\dagger$ (of q, given $\mathcal{J}$). Combining this with Theorem 4.8, we have the following result, announced in [NS96c].

Theorem 4.11 ([NS96d]) *There is a sub-subsequence L_{n_k} of the L_n's of Theorem 4.8 such that along some subsequence K_m of the K's,*

$$\textit{for } \nu\textit{-a.e. } \omega, \quad \kappa_{J(\omega)}^{K_m}((L_{n_k})) \to \kappa_{J(\omega)} \quad \textit{as } m \to \infty, \tag{4.32}$$

where κ_J is the same metastate as in Theorem 4.8.

Proof. This result follows from Proposition B.6 of Appendix B. □

We proceed, as in [NS96c], by giving a partial classification of the possible types of metastates κ_J which could occur. The simplest of these, and one which of course does occur for small β or for $d = 1$, is possibility 1) that κ_J is (for $\bar{\nu}$-a.e. $\mathcal{J}$) supported on a single pure Gibbs state $P = P_J$. This is the case, for example, if $\mathcal{N} = \mathcal{N}(d, \mu, \beta) = 1$ (e.g., for $\beta < \beta_c(d, \mu)$), since then $P_{J,L} \to P_J$ as $L \to \infty$. Another simple possibility 2) is that $\kappa = \kappa_J$ is supported on a single Gibbs state, which is a mixture of two distinct pure states:

$$\kappa = \delta_P, \quad P = \frac{1}{2}P' + \frac{1}{2}P'', \tag{4.33}$$

where $P' = P_J'$ and $P'' = P_J''$ are pure states that are global flips of each other. This would be the case, for example, according to the Fisher-Huse (FH) scaling picture discussed earlier, which predicts that (when $\mathcal{N}(d, \mu, \beta) \neq 1$) such a P_J' and P_J'' are the only pure states so that (by obvious spin flip symmetry arguments) $P_{J,L} \to \frac{1}{2}P_J' + \frac{1}{2}P_J''$ as $L \to \infty$. Of course, when $P_{J,L}$ converges to some P_J, there is no CSD and no need for a metastate description. When the metastate is supported on a single P_J, we will say that it is not dispersed.

A trivial sort of dispersal should occur in the FH picture if one replaces periodic boundary conditions by b.c.'s which break the spin-flip symmetry (but still do not depend on J) such as random b.c.'s (i.e., the boundary spins on $\partial\Lambda_L$ are i.i.d. symmetric random variables and independent of J) or more simply plus b.c.'s. Here, one expects the finite volume state to exhibit CSD, approximating P' for (roughly) half of the L's and P'' for the other half so that the metastate (for such a b.c.) should be $\frac{1}{2}\delta_{P'} + \frac{1}{2}\delta_{P''}$, which is quite different from (4.33). "Toy" versions of two-state CSD should also occur in some simpler contexts (see [En90], [NS92]) such as in the homogeneous Ising ferromagnet with random b.c.'s and in random field Ising models with periodic or free b.c.'s. We remark that metastates in random field Curie-Weiss models have been studied in [AmPZ92] and (along with metastates in Hopfield models) in [Ku96]. In particular, it has been shown in [Ku96] that the answer to (the analogue of) Research Problem 4.10 is negative; i.e., if subsequences of volumes are not taken, the convergence is valid only in distribution and not almost surely. Heuristically, one would expect the same behavior for the random field nearest-neighbor model.

Other sorts of dispersal could happen if there are more then two pure states, as in an SK picture. Roughly speaking, according to Theorem 4.11, a dispersed metastate describes the way in which, for large L,

$$P_{\mathcal{J},L} \approx \sum_{\alpha} W_{\mathcal{J},L}^{\alpha} P_{\mathcal{J}}^{\alpha}, \tag{4.34}$$

when there is CSD and $P_{\mathcal{J},L}$ does *not* converge to a single $P_{\mathcal{J}}$.

In possibilities 1 and 2 that we discussed above, most of the weight (represented by the $W_{\mathcal{J},L}^{\alpha}$'s) is concentrated on one or two pure states as $L \to \infty$. Two other possibilities (numbers 5 and 6, the latter of which includes the nonstandard SK picture) that we will soon discuss also have the property that most weight is concentrated on few states, even though there are (uncountably) many pure states. But first, we briefly mention two possibilities (very different from either FH or SK pictures) where the weights are shared more equitably. In both of these possibilities $\kappa_{\mathcal{J}}$ is supported on infinite volume Gibbs states Γ whose decomposition into pure states is continuous. Possibility 3) is that there is no dispersal and $\kappa_{\mathcal{J}}$ is supported on a single such Γ, while possibility 4) is that there is dispersal and $\kappa_{\mathcal{J}}$ is supported on multiple such Γ's. As there seems to be no particular reason to suppose that either of these possibilities occurs in the EA model, we proceed to the next possibility, which we find rather intriguing.

Possibility 5) is one in which there is CSD and dispersal among (uncountably) many Γ's, but each of these Γ's decomposes into just two pure states related by a global flip, like in possibility 2. This possibility, where $\mathcal{N} = \infty$ but for each large L one "only sees two pure states" (in the sense that $P_{\mathcal{J},L}$ is approximately a mixture of just two pure states) is intermediate between the scaling picture where $\mathcal{N} = 2$ and the next possibility (which leads to the nonstandard SK picture) where $\mathcal{N} = \infty$ and for large L one "sees many pure states".

Possibility 6) has, like possibility 5, nontrivial dispersal over many Γ's, but unlike possibility 5, its Γ's have a nontrivial decomposition into pure states — namely

$$\Gamma = \sum_{\alpha} W_{\Gamma}^{\alpha} P_{\mathcal{J}}^{\alpha} \; ; \tag{4.35}$$

this decomposition should be discrete (i.e., a countable sum, as indicated) but with many (in particular, more than two) nonzero weights W_{Γ}^{α}. As we shall see, the nonstandard SK picture will require that the dispersal of $\kappa_{\mathcal{J}}$ over Γ's of the form (4.35) is a dispersal not only over the weights W_{Γ}^{α} but also over the selection of the countably many pure states appearing in the discrete sum (from an uncountable family of all pure states for the given $\mathcal{J}$). To explain why that is so, we first need to discuss replicas and their overlaps within the metastate framework. (We remark, as previously discussed in the context of the standard SK picture, that, strictly speaking, it is the overlaps which must be discrete rather than the pure state decomposition.)

Replicas and overlaps are a way of probing the meaning of the approximate equation (4.34), our EA model replacement for the SK approximate equation (4.16). It is therefore natural to take replicas for fixed L and only afterward let $L \to \infty$. As emphasized by Guerra [Gu95], this order of operations could yield a different result than that obtained by first letting $L \to \infty$ and then taking replicas, as was done in the standard SK picture. In other words, rather than letting $L \to \infty$ first to obtain a single infinite volume state $P_\mathcal{J}$ and then defining replicas $s^1, s^2, s^3, \ldots$ by using the product measure $P_\mathcal{J}(s^1) \times P_\mathcal{J}(s^2) \times \ldots$, we will first take the product measure for finite volume, $P_{\mathcal{J},L}(s^1) \times P_{\mathcal{J},L}(s^2) \times \ldots$, and then let $L \to \infty$.

We have already encountered three (related) ways to consider the limit as $L \to \infty$ of $P_{\mathcal{J},L}$ (without replicas). These are: (a) to obtain $P_\mathcal{J}$ via the limit of the joint distribution $(\bar{\nu}(\mathcal{J}) \times P_{\mathcal{J},L}(s))$ of J and S (see Remark 4.9), (b) to obtain $\kappa_\mathcal{J}$ via the limit of the joint distribution of J and $P_{J,L}$ (see Theorem 4.8) and (c) to obtain $\kappa_\mathcal{J}$ via the limit for fixed $\mathcal{J}$ of the empirical distribution of $P_{\mathcal{J},L}$ (see Theorem 4.11). We want to see what happens to these three types of limits when replicas are taken first, i.e., when $P_{\mathcal{J},L}$ is replaced by the infinite product measure (on $\mathcal{S}_{\Lambda_L} \times \mathcal{S}_{\Lambda_L} \times \ldots$),

$$P^\infty_{\mathcal{J},L}(s^1, s^2, \ldots) = P_{\mathcal{J},L}(s^1) \times P_{\mathcal{J},L}(s^2) \times \ldots, \tag{4.36}$$

which is the finite volume Gibbs state for (arbitrarily many) replicas. The next proposition (the first part of which is explicit in [NS96c] with the rest implicit) gives the answer, which is that $\kappa_\mathcal{J}(\Gamma)$ is replaced by $\kappa^\infty_\mathcal{J}$, the probability measure on $\Omega^1 \times \Omega^1 \times \ldots$, supported on elements of the form $\Gamma \times \Gamma \times \ldots$, with Γ distributed by $\kappa_\mathcal{J}$. Convergence of measures or random variables in the proposition denotes, as usual, (weak) convergence of finite dimensional distributions.

Proposition 4.12 ([NS96d]) *Let L_n, n_k and K_m be as in Theorems 4.8 and 4.11. Then*

$$\bar{\nu}(\mathcal{J}) \times P^\infty_{\mathcal{J},L_n} \to \bar{\nu}(\mathcal{J}) \times P^\infty_\mathcal{J} \qquad as \ n \to \infty, \tag{4.37}$$

where $P^\infty_\mathcal{J}$ is the measure on S^∞,

$$P^\infty_\mathcal{J}(s^1, s^2, \ldots) = \int \left[\Gamma(s^1) \times \Gamma(s^2) \times \ldots \right] d\kappa_\mathcal{J}(\Gamma); \tag{4.38}$$

$$(J, P^\infty_{J,L_n}) \to \bar{\nu}(\mathcal{J}) \times \kappa^\infty_\mathcal{J} \qquad as \ n \to \infty; \tag{4.39}$$

and finally, defining the empirical distribution for replicas to be the measure on $\Omega^1 \times \Omega^1 \times \ldots$,

$$\kappa^{K,\infty}_J((L_n)) = \frac{1}{K} \sum_{k=1}^{K} \delta_{(\vec{Q}^{(L_k)}, \vec{Q}^{(L_k)}, \ldots)}, \tag{4.40}$$

we have

$$\text{for } \nu\text{-a.e. } \omega, \quad \kappa_{J(\omega)}^{K_m,\infty}\left((L_{n_k})\right) \to \kappa_{J(\omega)}^{\infty} \qquad \text{as } m \to \infty. \qquad (4.41)$$

Proof. We sketch the proofs in the reverse order of the proposition. The last claim (4.41) easily reduces to the following fact, whose proof we leave to the reader: if $\vec{Q}_l$ is a sequence of random vectors whose sequence of distributions (on $\mathbb{R}^M$, for some M) converges as $l \to \infty$ to some ρ, then the sequence of distributions of $(\vec{Q}_l, \vec{Q}_l, \ldots, \vec{Q}_l)$ (on $(\mathbb{R}^M)^k)$) converges to $\rho \times \rho \times \ldots \times \rho$. Similarly, the middle claim (4.39) reduces to: if $(\vec{J}_l, \vec{Q}_l)$ converges in distribution to some $\rho^\dagger$ whose marginal (of $\vec{q}$, given $\vec{J}$) is $\rho_{\vec{J}}$, then $(\vec{J}_l, \vec{Q}_l, \ldots, \vec{Q}_l)$ converges to a measure whose marginal (of $(\vec{q}, \ldots, \vec{q})$, given $\vec{J}$) is $\rho_{\vec{J}} \times \ldots \times \rho_{\vec{J}}$. The first claim (4.37) follows from (4.39) by the replica version of Remark 4.9. This is essentially the same as arguing directly from Theorem 4.8 by using "metacorrelations", as in [NS96c]. That argument makes clear that replicas are already implicit in the very notion of metastates, since the natural generalization of (4.27) beyond functions linear in q replaces $\langle S_A \rangle_{J,L_n}$ by the metacorrelation

$$\langle S_{A_1} \rangle_{J,L_n} \cdots \langle S_{A_m} \rangle_{J,L_n} = \langle S_{A_1}^1 \cdots S_{A_m}^m \rangle_{J,L_n}^\infty, \qquad (4.42)$$

where the latter bracket denotes the expectation with respect to P_{J,L_n}^∞.   □

Remark 4.13 There are two conceptually important points hidden in the last proposition. One, which is the essential content of (4.41), is that replicas must be described in the infinite volume limit by $\Gamma \times \Gamma \times \ldots$ with Γ distributed by κ_J because replicas in finite volume are taken from the *same* L and κ_J (according to Theorem 4.11) describes the sampling of states as L varies. The second, emphasized by Guerra [Gu95], is that in the infinite volume replica measure $P_{\vec{J}}^\infty$, the replicas need not be independent, even though they of course are independent (by construction) in the finite volume measure $P_{\vec{J},L}^\infty$. Indeed, from the form of (4.38), we see that they will only be independent if κ_J is non-dispersed. In [NS96c], this lack of dependence when the metastate is dispersed is called replica non-independence (RNI). It is equivalent to the non-interchangeability of the operations of taking replicas and letting $L \to \infty$, since done in the other order (as in the standard SK picture) the measure obtained is

$$P_J(s^1) \times P_J(s^2) \times \ldots = \int \Gamma_1(s^1) d\kappa_J(\Gamma_1) \times \int \Gamma_2(s^2) d\kappa_J(\Gamma_2) \times \ldots, \qquad (4.43)$$

which does not agree with $P_{\vec{J}}^\infty$ unless κ_J is non-dispersed.

Now that we have seen that replicas are described by $\kappa_{\vec{J}}^\infty$, it is fairly clear what the appropriate replacements are for the overlap R_J and overlap distribution $\mathcal{P}_J$ as defined (in the standard SK picture) by using the two-replica product measure $P_J \times P_J$. Namely, on $\mathcal{S} \times \mathcal{S}$ with product measure $\Gamma \times \Gamma$, we define the random variable

$$R_\Gamma = \lim_{L \to \infty} |\Lambda_L|^{-1} \sum_{x \in \Lambda_L} s_x^1 s_x^2, \qquad (4.44)$$

and define $\mathcal{P}_\Gamma$ to be (the formal density of) its distribution.

The nature of $\mathcal{P}_\Gamma$ depends of course on the nature of Γ, and more particularly on its pure state decomposition. Thus if Γ is a pure state (and the limit in (4.44) exists), R_Γ will be the constant r_Γ in which $s_x^1 s_x^2$ in (4.44) is replaced by the expectation $\langle s_x \rangle_\Gamma^2$; if Γ is pure and flip invariant then $\langle s_x \rangle_\Gamma \equiv 0$ and $\mathcal{P}_\Gamma = \delta(r)$. If Γ has the decomposition $\Gamma = \frac{1}{2}P' + \frac{1}{2}P''$ with P' and P'' related by a global flip (as in possibilities 2 and 5 discussed above), then $\mathcal{P}_\Gamma = \frac{1}{2}\delta(r - r_0) + \frac{1}{2}\delta(r + r_0)$, where $r_0 = r_{P'} = r_{P''}$. If Γ is as in (4.35) of possibility 6, then $\mathcal{P}_\Gamma$ has the SK type form (see (4.18)),

$$\mathcal{P}_\Gamma = \sum_\alpha \sum_\gamma W_\Gamma^\alpha W_\Gamma^\gamma \delta(r - r_{\alpha\gamma}). \qquad (4.45)$$

It is important to note that our analysis of the relation between the state Γ and the overlap distribution $\mathcal{P}_\Gamma$ is based on the definition (4.44) for the overlap. The limit in (4.44) is taken within an already infinite system, corresponding physically to taking the overlap in a box which is large but much smaller than the overall system size. Were one to take the overlap box and system size the same, one might, in principle, obtain a different limiting distribution. For example, Huse and Fisher [HuF87] noted that in the $d = 2$ homogeneous Ising ferromagnet, such a limit (with antiperiodic b.c.'s) yields a *continuous* overlap distribution (rather than $\frac{1}{2}\delta(r - r_0) + \frac{1}{2}\delta(r + r_0)$) despite the presence of only two pure states.

In the metastate approach, even with $\mathcal{J}$ fixed, Γ should itself be regarded as random with the distribution $\kappa_\mathcal{J}(\Gamma)$. Thus $\mathcal{P}_\Gamma$ should be regarded as random with the distribution induced on it by $\kappa_\mathcal{J}$. In possibilities 1 and 2 (and 3), $\kappa_\mathcal{J}$ is non-dispersed, so $\mathcal{P}_\Gamma$ is non-random. In possibility 5, $\kappa_\mathcal{J}$ is dispersed, but if we postulate (as in the SK pictures) that $r_{\alpha\alpha}$ (see (4.17)) is the same ($= r_{EA}$) for all pure states $P_\mathcal{J}^\alpha$, then $\mathcal{P}_\Gamma$ is again non-random. In possibility 6, $\kappa_\mathcal{J}$ is dispersed and $\mathcal{P}_\Gamma$ should really be random, i.e., should depend on Γ since the W_Γ^α's and the selection of the $r_{\alpha\gamma}$'s that have nonzero weights should depend on Γ.

This dependence on Γ (for fixed $\mathcal{J}$) is crucial for the nonstandard SK picture because there can be no dependence on $\mathcal{J}$. More precisely, if one considers the distribution of the random $\mathcal{P}_\Gamma$ (that it inherits from $\kappa_\mathcal{J}$) as a function of $\mathcal{J}$, there is a translation invariance/ergodicity argument, essentially the same as in Proposition 4.6 and the discussion preceding it, which shows that that distribution is self-averaged. Such a self-averaging property killed the standard SK picture. It does not kill the nonstandard picture (at least, not on its own) because there *dependence on $\mathcal{J}$ is replaced by dependence on Γ for fixed $\mathcal{J}$*. In particular $\bar{\mathcal{P}} = \int \mathcal{P}_\mathcal{J} d\bar{\nu}(\mathcal{J})$ is now replaced by

$$\tilde{\mathcal{P}} = \int \mathcal{P}_\Gamma d\kappa_\mathcal{J}(\Gamma). \qquad (4.46)$$

The nonstandard SK picture thus includes the following requirements within the context of possibility 6: (1) $\mathcal{P}_\Gamma$ does depend on Γ as Γ varies according to $\kappa_\mathcal{J}$; (2) $\mathcal{P}_\Gamma$ is (for $\kappa_\mathcal{J}$-a.e. Γ) a sum of (countably) infinitely many delta functions; (3)

$\tilde{\mathcal{P}}$ is continuous (except for two δ-functions at $\pm r_{EA}$, whose weights add up to less than one); and (4) ultrametricity of $\{r_{\alpha\gamma} : W_\Gamma^\alpha W_\Gamma^\gamma \neq 0\}$ (for $\kappa_{\mathcal{J}}$-a.e. Γ).

Before concluding, we point out that replica symmetry breaking (RSB), which plays an important role in the usual analyses of the SK model, has a very natural interpretation within the metastate framework. Replica symmetry just refers to permutation invariance among the replica labels. To be broken, there must be something which distinguishes between the replicas, such as nontrivial $q^{\alpha\gamma}$'s with α and γ coming from different replicas, or more generally different pure states coming from different replicas. Thus we consider RSB to occur whenever the Γ's in the support of the metastate are not pure states. Thus when $\Gamma \times \Gamma \times \ldots$ is decomposed into pure states, different α's and γ's will appear for different replicas. In the physics literature, it is conventional to regard situations such as possibility 2, where the replicas appear in pure states which only differ by a global flip, as a trivial example of RSB. Thus we would describe possibility 2 as having no dispersal of the metastate and only trivial RSB while possibility 5 has dispersal of the metastate (along with CSD and RNI) but still only trivial RSB. The nonstandard SK picture has both dispersal of the metastate and nontrivial RSB.

Should there be an affirmative answer to Research Problem 4.5, i.e., should it be shown that the number of pure states (sometimes) exceeds 2, it would remain to determine whether the many states are "seen" in $P_{\mathcal{J},L}$ for large L and if so, how. The nonstandard SK picture is only one of a number of ways in which the many states might be seen, but it appears to be the only way in which the essential features (suitably reinterpreted) of Parisi's mean-field analysis would be present. We leave as part of our last research problem the question, is the nonstandard SK picture valid for any d, μ and β?

Research Problem 4.14 *Determine whether for some d, μ and β, the metastate $\kappa_{\mathcal{J}}$ is (for $\bar{\nu}$-a.e. $\mathcal{J}$) supported on (a) many Γ's, (b) Γ's which decompose into more than two pure states, (c) Γ's which satisfy the other requirements of the nonstandard SK picture.*

Appendix A

Infinite Geodesics and Measurability

As in the proof of Proposition 1.4 in Chapter 1, we take each $\tau(e^*)$ to be the coordinate function ω_{e^*} on $(\Omega, \mathcal{F}, \nu)$, the product over $\mathbb{E}^{d*}$ of $(\mathbb{R}, \mathcal{B}, \mu^*)$, where $\mathcal{B}$ is the Borel σ-field on $\mathbb{R}$ and μ^* is some probability measure on $(\mathbb{R}, \mathcal{B})$. We want to show that certain subsets of Ω, defined in Chapter 1 in terms of finite and infinite geodesics, are indeed events - i.e., that they do belong to the σ-field $\mathcal{F}$. Our main purpose here is to consider sets whose definition involves the existence or nonexistence of $\hat{x}$-unigeodesics (for a specified deterministic $\hat{x}$). These are not so obviously measurable. In fact, we will only give a complete proof of measurability for $d = 2$, which suffices for the purposes of Chapter 1. Towards the end of this appendix, we give some partial results for general d.

We begin with some simpler subsets of Ω, not involving infinite geodesics. By a path r, we will always mean a site-self-avoiding nearest neighbor path on $\mathbb{Z}^{d*}$, and for finite r, $T(r)$ denotes $\sum_{e^* \in r} \tau(e^*)$. If the first site of r is u and the last is v (or vice-versa) we say r connects u and v. The following sets are all easily seen to be in $\mathcal{F}$ (and they have ν-probability one, when μ is a continuous measure on $(0, \infty)$):

$$\{\omega \in \Omega : \ \forall e^* \in \mathbb{E}^{d*}, \ \tau(e^*) > 0\}, \tag{A.1}$$

$$\{\omega \in \Omega : \ \forall u \neq v \text{ in } \mathbb{Z}^d \text{ and } r \neq r' \text{ connecting } u \text{ and } v, \ T(r) \neq T(r')\}, \tag{A.2}$$

$$\{\omega \in \Omega : \ \forall u \neq v \text{ in } \mathbb{Z}^d, \ \exists r \text{ connecting } u \text{ and } v \text{ such that} \tag{A.3}$$
$$\forall r' \neq r \text{ connecting } u \text{ and } v, \ T(r) < T(r')\}.$$

We define Ω_0 to be the intersection of these events and henceforth only consider subsets of Ω_0.

On Ω_0, there is, for every u, v, a unique finite geodesic, $M(u, v)$, which is itself measurable; e.g., for every finite path $\tilde{r}$ connecting u and v,

$$\{M(u, v) = \tilde{r}\} = \{\omega \in \Omega_0 : \ \forall r' \neq \tilde{r} \text{ connecting } u \text{ and } v, \ T(\tilde{r}) < T(r')\} \in \mathcal{F}. \tag{A.4}$$

Furthermore, the $M(u, v)$'s have the property that if u', v' are vertices in $M(u, v)$, then $M(u', v')$ is a subpath of $M(u, v)$. This property leads to two key constructions in Chapter 1, which are the spanning tree $R(u)$ and its subgraph $\hat{R}(u)$. $R(u)$, being

65

the union of the $M(u, v)$'s as v varies over the countably many sites in $\mathbb{Z}^{d*}$, is clearly measurable (e.g., $\{e^* \in R(u)\} \in \mathcal{F}$). A unigeodesic starting at u is a semi-infinite path starting at u which is a subgraph of $R(u)$. To see that $\hat{R}(u)$, the union of all unigeodesics starting at u, is measurable, note that for any finite path $\tilde{r}$ starting at u,

$$
\begin{aligned}
A_{\tilde{r}} &\equiv \{\omega \in \Omega_0 : \exists \text{ a unigeodesic whose initial segment is } \tilde{r}\} \\
&= \{\omega \in \Omega_0 : \exists \text{ infinitely many } v\text{'s such that } M(u, v) \qquad (A.5) \\
&\qquad \text{begins with initial segment } \tilde{r}\} \in \mathcal{F}.
\end{aligned}
$$

When working with a fixed $R(u)$, here and elsewhere, we regard $M(u, v)$ as ordered so that it starts at u and ends at v.

We now restrict attention to $d = 2$, but we will later return to general d. For $d = 2$, two other key constructions in Chapter 1 are $\hat{r}^+(u, v)$ and $\hat{r}^-(u, v)$. More generally, for a finite path $\tilde{r}$ starting at u, we similarly define $\hat{r}^+(\tilde{r}) = (u_0^+(\tilde{r}), u_1^+(\tilde{r}), \ldots)$ and $\hat{r}^-(\tilde{r}) = (u_0^-(\tilde{r}), u_1^-(\tilde{r}), \ldots)$ as the most "counterclockwise" and "clockwise" paths in $\hat{R}(u)$ with initial segment $\tilde{r}$; these are undefined unless $\tilde{r}$ is in $\hat{R}(u)$. Since $\hat{R}(u)$ is measurable and each step along $\hat{r}^\pm(\tilde{r})$ is defined by looking at a finite portion of $\hat{R}(u)$, it is clear that $\hat{r}^\pm(\tilde{r})$ is also measurable.

For a finite or semi-infinite path $r = (u_0, u_1, u_2, \ldots)$ starting at u_0, we define $\theta_i = \theta_i[r]$ in $(-\infty, \infty)$ for $i \geq 1$, such that the actual angle in $[-\pi, \pi]$ from $u_1 - u_0$ to $u_i - u_0$ equals $\theta_i[r]$ mod 2π, as follows. Draw the polygonal path on the Riemann surface of $\log z$ starting with $z_1 = 1$ on the principal sheet and adding $u_{i+1} - u_i$ to z_i at each subsequent step; then define θ_i as the "angle" of z_i ($-\pi < \theta_i \leq \pi$ on the principal sheet, $\pi < \theta_i \leq 3\pi$ on the next sheet above, etc.). For r semi-infinite, we also define $\bar{\theta}[r]$ and $\underline{\theta}[r]$ as the lim sup and the lim inf of $\theta_i[r]$ as $i \to \infty$. Define $k_0 \in \{-1, 0, 1, 2\}$ by $u_1 - u_0 = (\cos(k_0\pi/2), \sin(k_0\pi/2))$. It should be clear that

$$
\lim_{i \to \infty} \frac{u_i}{||u_i||} = \hat{x} = (\cos\theta, \sin\theta) \quad \Longleftrightarrow \quad \bar{\theta}[r] = \underline{\theta}[r] = \theta + 2k\pi - k_0\pi/2 \qquad (A.6)
$$

for some $k \in \mathbb{Z}$. For our countably many special paths $\hat{r}^\pm(\tilde{r})$, we use the notation

$$
\bar{\theta}^\pm(\tilde{r}) = \bar{\theta}[\hat{r}^\pm(\tilde{r})] \quad \text{and} \quad \underline{\theta}^\pm(\tilde{r}) = \underline{\theta}[\hat{r}^\pm(\tilde{r})]. \qquad (A.7)
$$

Since $\hat{r}^\pm(\tilde{r})$ is measurable, so is $u_i^\pm(\tilde{r})$, and thus so is $\bar{\theta}^\pm(\tilde{r})$ and $\underline{\theta}^\pm(\tilde{r})$. Of course, these random variables are undefined on certain events (the complement of $A_{\tilde{r}}$) and can be $+\infty$ or $-\infty$, even when defined.

Using these special random variables, we will prove first that

$$
A_{(u,v)}(\hat{x}) = \{\omega \in \Omega_0 : \exists \text{ an } \hat{x}\text{-unigeodesic starting with } (u, v)\} \qquad (A.8)
$$

is in $\mathcal{F}$. Once this is shown, it will follow that the intersection of Ω_0 and $D_U(\hat{x})$, defined in (1.10), is also in $\mathcal{F}$. This is so because

$$
\Omega_0 \cap D_U(\hat{x}) = \Omega_0 \setminus \bigcup_{u \in \mathbb{Z}^{2*}} \bigcup{}' \left[A_{(u,v_1)}(\hat{x}) \cap A_{(u,v_2)}(\hat{x}) \right], \qquad (A.9)
$$

where $\bigcup'$ denotes the union over all distinct v_1 and v_2 which are both nearest neighbors of u. The first main result of this appendix is the following proposition. After we give its proof, we will see that related arguments show that (the intersections of Ω_0 with) other sets considered in Chapter 1, such as $\{\exists \text{ disjoint } \hat{x}\text{-unigeodesics }\}$ or $\{\exists \text{ an } (\hat{x}, -\hat{x})\text{-bigeodesic }\}$, are also in $\mathcal{F}$.

Proposition A.1 *For $d = 2$, any $\{u, v\} \in \mathbb{Z}^{2*}$ and any unit vector $\hat{x}$ in $\mathbb{R}^2$, the set $A_{(u,v)}(\hat{x})$ defined in (A.8) is in $\mathcal{F}$. Hence, so is $\Omega_0 \cap D_U(\hat{x})$.*

The proof of this proposition is based on several lemmas. In each lemma, $\omega \in \Omega_0$.

Lemma A.2 *Suppose $r = (u_0, u_1, \ldots)$ and $r' = (u'_0, u'_1, \ldots)$ are unigeodesics such that for some $m \geq 1$, $u_i = u'_i$ for $0 \leq i \leq m$ while $\theta_{m+1}[r] < \theta_{m+1}[r']$. Then*

$$\bar{\theta}[r'] - 2\pi \leq \bar{\theta}[r] \leq \bar{\theta}[r'] \quad and \quad \underline{\theta}[r'] - 2\pi \leq \underline{\theta}[r] \leq \underline{\theta}[r']. \tag{A.10}$$

Proof. Because $\omega \in \Omega_0$, once the geodesics r and r' bifurcate from each other, they must remain disjoint. We next give heuristic arguments, followed by more precise ones (the latter suggested by Y. Eliashberg). After bifurcation, r is located "clockwise" from r'; this yields $\bar{\theta}[r] \leq \bar{\theta}[r']$ and $\underline{\theta}[r] \leq \underline{\theta}[r']$. As to the other inequalities, if any were violated, it would mean that r had gone an entire extra clockwise revolution compared to r', which is geometrically impossible in two dimensions.

Now, to be more precise, we consider the two polygonal paths (starting from $z_1 = 1$ on the principal sheet) on the $\log z$ Riemann surface corresponding to r and r' (starting with u_1). Each point z on each path has a well defined $\rho(z) = |z| \in [1, \infty)$ and $\theta(z) \in (-\infty, \infty)$. Denote by γ and γ' the two loci $\{(\rho(z), \theta(z))\}$ from r and r'. γ and γ' are both non-self-intersecting. They coincide until $z = z_m$ and then are disjoint. Indeed, after bifurcation, also $(\rho', \theta') \neq (\rho, \theta + 2\pi k)$ for any integer k, because the original paths in $\mathbb{Z}^{2*}$ are then disjoint. The intersection of γ or γ' with any half space, $\rho \leq \text{const.}$, is bounded. γ' divides $[1, \infty) \times (-\infty, \infty)$ into upper and lower portions and γ is in the lower portion. In particular, the maximum (resp., minimum) θ in the intersection of any vertical line, $\rho = \lambda$, with γ', which we denote by $\bar{\Theta}'(\lambda)$ (resp., $\underline{\Theta}'(\lambda)$) is $\geq$ the corresponding $\bar{\Theta}(\lambda)$ (resp., $\underline{\Theta}(\lambda)$) for γ. But $\bar{\theta}[r'] = \limsup_{\lambda \to \infty} \bar{\Theta}'(\lambda)$ and analogous identities apply to $\underline{\theta}[r']$, $\bar{\theta}[r]$ and $\underline{\theta}[r]$. This yields $\bar{\theta}[r] \leq \bar{\theta}[r']$ and $\underline{\theta}[r] \leq \underline{\theta}[r']$. Finally, let γ'' be γ' with its θ coordinate reduced by 2π. Then γ'' also divides $[1, \infty) \times (-\infty, \infty)$ into two portions, but now γ is in the upper portion. This yields the remaining inequalities. $\square$

Lemma A.3 *Suppose $\exists$ an $\hat{x}$-unigeodesic r^* starting with (u, v). Then $\bar{\theta}^{\pm}(u, v)$ and $\underline{\theta}^{\pm}(u, v)$ are finite and*

$$\begin{aligned} 0 &\leq \underline{\theta}^+(u, v) - \underline{\theta}^-(u, v) \leq 2\pi, \\ 0 &\leq \bar{\theta}^+(u, v) - \bar{\theta}^-(u, v) \leq 2\pi. \end{aligned} \tag{A.11}$$

Proof. If $\hat{r}^+(u, v) = r^*$, then of course $\bar{\theta}^+(u, v)$ and $\underline{\theta}^+(u, v)$ are finite (and equal). Otherwise, applying Lemma A.2 with $r' = \hat{r}^+(u, v)$ and $r = r^*$ gives their finiteness. Similarly, if $\hat{r}^-(u, v) \neq r^*$, then applying Lemma A.2 with $r' = r^*$ and $r = \hat{r}^-(u, v)$ gives the finiteness of $\bar{\theta}^-(u, v)$ and $\underline{\theta}^-(u, v)$. Finally, applying Lemma A.2 with $r' = \hat{r}^+(u, v)$ and $r = \hat{r}^-(u, v)$ (assuming these are not equal) gives (A.11). □

The next two lemmas relate $\hat{x}$-unigeodesics to the random variables $\bar{\theta}^\pm(\tilde{r})$ and $\underline{\theta}^\pm(\tilde{r})$. In each lemma $\tilde{r}$ denotes a finite path (with two or more vertices) which starts with (u_0, u_1), k_0 is as in (A.6) and, as in the previous lemmas, $\omega \in \Omega_0$.

Lemma A.4 *If $\exists$ an $\hat{x}$-unigeodesic r^* starting with $\tilde{r}$, then $\hat{x} = (\cos\theta, \sin\theta)$ for some θ in the closed interval,*

$$I(\tilde{r}) = [\bar{\theta}^-(\tilde{r}), \underline{\theta}^+(\tilde{r})] + k_0\pi/2. \tag{A.12}$$

Proof. From Lemma A.2 and the definitions of $\hat{r}^\pm(\tilde{r})$, it follows that

$$\bar{\theta}^-(\tilde{r}) \leq \bar{\theta}(r^*), \quad \underline{\theta}(r^*) \leq \underline{\theta}^+(\tilde{r}). \tag{A.13}$$

Since r^* is an $\hat{x}$-unigeodesic, we have from (A.6) that $\bar{\theta}(r^*) = \underline{\theta}(r^*) = \theta^*$ with $\hat{x} = (\cos\theta, \sin\theta)$ for $\theta = \theta^* + k_0\pi/2$. By (A.13), $\theta^* \in [\bar{\theta}^-(\tilde{r}), \underline{\theta}^+(\tilde{r})]$, which completes the proof. □

Lemma A.5 *Suppose $\tilde{r} = (u_0, u_1, \ldots, u_m)$ is in $\hat{R}(u_0)$ and there are exactly $\tilde{j}$ ($= 1$ or 2 or 3) "one-step extensions" of $\tilde{r}$ in $\hat{R}(u_0)$, which we denote $\tilde{r}^{(j)} = (u_0, u_1, \ldots, u_m, u_{m+1}^{(j)})$, for $1 \leq j \leq \tilde{j}$. Suppose further that for each j and each choice of $+$ or $-$, $\hat{r}^\pm(\tilde{r}^{(j)})$ is not an $\hat{x}$-unigeodesic. If $\exists$ an $\hat{x}$-unigeodesic r^* starting with $\tilde{r}$, then there is exactly one value j^* for j such that the closed interval $I(\tilde{r}^{(j)})$ (as defined in (A.12)) intersects $\{\theta \in \mathbb{R} : \hat{x} = (\cos\theta, \sin\theta)\}$; furthermore, r^* starts with $\tilde{r}^{(j^*)}$.*

Proof. We assume $\tilde{j} \neq 1$ since otherwise the conclusions are trivial. By relabelling, we may also assume that $\theta_{m+1}(\tilde{r}^{(j)})$ is (strictly) increasing in j. Then, by Lemmas A.2 and A.3, the intervals $I(\tilde{r}^{(j)}) = [L_j, R_j]$ are finite and ordered, in the sense that $R_j \leq L_{j+1}$ for $j < \tilde{j}$. Furthermore $R_{\tilde{j}} - L_1 \leq 2\pi$. Since r^* starts with some $\tilde{r}^{(j)}$, the conclusions all follow from Lemma A.4 unless one of the following two cases occurs:

1. for some $j < \tilde{j}$, $R_j = L_{j+1} = \theta$ with $\hat{x} = (\cos\theta, \sin\theta)$.

2. $R_{\tilde{j}} = L_1 + 2\pi = \theta$ with $\hat{x} = (\cos\theta, \sin\theta)$.

But if case (1) occurs, then by Lemma A.2

$$\underline{\theta}^+(\tilde{r}^{(j)}) = \bar{\theta}^+(\tilde{r}^{(j)}) = \underline{\theta}^-(\tilde{r}^{(j+1)}) = \bar{\theta}^-(\tilde{r}^{(j+1)}) = \theta, \tag{A.14}$$

with $\hat{x} = (\cos\theta, \sin\theta)$, which means that both $\hat{r}^+(\tilde{r}^{(j)})$ and $\hat{r}^-(\tilde{r}^{(j+1)})$ are $\hat{x}$-unigeodesics, contradicting an hypothesis of the lemma. Similarly, if case (2) occurs, then by Lemma A.2,

$$\underline{\theta}^+(\tilde{r}^{(\tilde{j})}) = \bar{\theta}^+(\tilde{r}^{(\tilde{j})}) = \underline{\theta}^-(\tilde{r}^{(1)}) + 2\pi = \bar{\theta}^-(\tilde{r}^{(1)}) + 2\pi = \theta, \quad (A.15)$$

with $\hat{x} = (\cos\theta, \sin\theta)$, so that $\hat{r}^+(\tilde{r}^{(\tilde{j})})$ and $\hat{r}^-(\tilde{r}^{(1)})$ are $\hat{x}$-unigeodesics, which is again a contradiction. The proof is now complete. □

Proof of Proposition A.1. We set $u_0 = u$, $u_1 = v$ and k_0 as in (A.6). For any finite path $\tilde{r}$, $\bar{\theta}^\pm(\tilde{r})$ and $\underline{\theta}^\pm(\tilde{r})$ are measurable and so, by (A.6), are

$$B_{\tilde{r}}^\pm(\hat{x}) \quad = \quad \left\{ \omega \in \Omega_0 : \ \hat{r}^\pm(\tilde{r}) \text{ exists and is an } \hat{x}\text{-unigeodesic} \right\} \quad (A.16)$$

and

$$B_{(u,v)}(\hat{x}) \quad = \quad \{\exists \text{ a finite path } \tilde{r} \text{ starting with } (u,v) \text{ such}$$
$$\text{that either } B_{\tilde{r}}^+(\hat{x}) \text{ or } B_{\tilde{r}}^-(\hat{x}) \text{ (or both) occur} \} . \quad (A.17)$$

We next give an algorithm, which will define, on a subset of Ω_0, a unigeodesic, $r^{\hat{x}}(u,v) = (u_0^{\hat{x}}(u,v), u_1^{\hat{x}}(u,v), \ldots)$. Each $u_i^{\hat{x}}(u,v)$ will be measurable because it will be determined in terms of a finite portion of $\hat{R}(u)$ and finitely many $\bar{\theta}^\pm(\tilde{r})$'s and $\underline{\theta}^\pm(\tilde{r})$'s.

The algorithm is defined inductively as follows. The zeroth vertex $u_0^{\hat{x}} = u_0^{\hat{x}}(u,v)$ is taken to be u. Given $r_m = (u_0^{\hat{x}}, \ldots, u_m^{\hat{x}})$, $u_{m+1}^{\hat{x}}$ is chosen so that r_{m+1} is a path in $\hat{R}(u)$ starting with (u,v) such that the closed interval $I(r_{m+1})$, defined in (A.12), intersects $\{\theta \in \mathbb{R} : \ \hat{x} = (\cos\theta, \sin\theta)\}$; if no such $u_{m+1}^{\hat{x}}$ exists or if more than one exists, then $u_i^{\hat{x}}$ is undefined for $i > m$. We will say $r^{\hat{x}}(u,v)$ exists if $u_i^{\hat{x}}(u,v)$ is defined for every i. It should be clear that

$$B_{(u,v)}^*(\hat{x}) = \left\{ \omega \in \Omega_0 : \ r^{\hat{x}}(u,v) \text{ exists and is an } \hat{x}\text{-unigeodesic} \right\} \quad (A.18)$$

is measurable (i.e., is in $\mathcal{F}$). If it were the case that

$$A_{(u,v)}(\hat{x}) = B_{(u,v)}(\hat{x}) \cup B_{(u,v)}^*(\hat{x}), \quad (A.19)$$

then $A_{(u,v)}(\hat{x})$ would be in $\mathcal{F}$ as the proposition claims. Indeed, the inclusion of the RHS of (A.19) in the LHS is trivial, while Lemma A.5 implies that on $A_{(u,v)}(\hat{x}) \setminus B_{(u,v)}(\hat{x})$, the geodesic $r^{\hat{x}}(u,v)$ exists and is the unique $\hat{x}$-unigeodesic starting with (u,v). Thus

$$A_{(u,v)}(\hat{x}) \setminus B_{(u,v)}(\hat{x}) \subseteq B_{(u,v)}^*(\hat{x}), \quad (A.20)$$

which gives the opposite inclusion, yielding (A.19) and completing the proof of Proposition A.1. □

Proposition A.6 *For $d = 2$, any u, u' in $\mathbb{Z}^{2*}$ and any unit vectors $\hat{x}$ and $\hat{y}$ in $\mathbb{R}^2$, the sets*

$$A^*_{u,u'}(\hat{x}) \quad = \quad \{\omega \in \Omega_0 : \exists \text{ vertex-disjoint } \hat{x}\text{-unigeodesics} \qquad (A.21)$$
$$\text{starting from } u \text{ and } u'\}$$

and

$$\tilde{A}_u(\hat{x}, \hat{y}) \quad = \quad \{\omega \in \Omega_0 : \exists \text{ an } (\hat{x}, \hat{y})\text{-bigeodesic passing through } u\} \quad (A.22)$$

are in $\mathcal{F}$. Hence, so are the intersections of Ω_0 with $\{\exists \text{ disjoint } \hat{x}\text{-unigeodesics }\}$ and with $\{\exists \text{ an } (\hat{x}, \hat{y})\text{-bigeodesic }\}$.

Proof. We observed in the proof of Proposition A.1 that on $A_{(u,v)}(\hat{x}) \setminus B_{(u,v)}(\hat{x})$, $r^{\hat{x}}(u,v)$ is the unique $\hat{x}$-unigeodesic starting with (u,v). Thus $A^*_{u,u'}(\hat{x})$ is in $\mathcal{F}$ since it is the countable union of the events of the form

$$\{\omega \in \Omega_0 : r^*_u \text{ and } r^*_{u'} \text{ exist, are } \hat{x}\text{-unigeodesics and are disjoint }\}, \qquad (A.23)$$

where r^*_u denotes either $r^{\hat{x}}(u,v)$ for some nearest neighbor of u or $\hat{r}^+(\tilde{r})$ or $\hat{r}^-(\tilde{r})$ for some finite path $\tilde{r}$ starting from u, and similarly for $r^*_{u'}$. Analogously, $\tilde{A}_u(\hat{x}, \hat{y})$ is in $\mathcal{F}$ since it is the countable union of the events of the form

$$\{\omega \in \Omega_0 : r^*_u = (u, u^1_1, u^1_2, \ldots) \text{ exists and is an } \hat{x}\text{-unigeodesic},$$
$$r^{**}_u = (u, u^2_1, u^2_2, \ldots) \text{ exists and is a } \hat{y}\text{-unigeodesic}, \qquad (A.24)$$
$$\text{and for every } m \geq 1, \ (u^2_m, \ldots, u^2_1, u, u^1_1, \ldots, u^1_m) = M(u^2_m, u^1_m)\},$$

where r^*_u is as in (A.23) and r^{**}_u is the same except with $\hat{y}$ replacing $\hat{x}$. This completes the proof. □

We now return to general d (≥ 2). For nonzero x and y in $\mathbb{R}^d$, we define $\theta(x,y)$ to be the angle in $[0, \pi]$ such that $\cos(\theta(x,y)) = (x,y)/(\|x\| \cdot \|y\|)$, where (x,y) is the standard inner product. We also define for $\delta \geq 0$, $\mathcal{C}(x, \delta)$ to be the cone, $\{y \in \mathbb{R}^d : \theta(x,y) \leq \delta\}$. Let $h(L)$ be a (deterministic) function from $(0, \infty)$ to $(0, \infty)$ such that $h(L) \to 0$ as $L \to \infty$; e.g.,

$$h(L) = L^{-(1/4-\varepsilon)} \quad \text{with } 0 < \varepsilon < 1/4. \qquad (A.25)$$

Following [N95], we will say the spanning tree $R(u)$ is h-straight if: for all but finitely many $v \in \mathbb{Z}^{d*} \setminus \{u\}$, every v', such that $M(u,v')$ passes through v, has $\theta(v - u, v' - u) \leq h(\|v - u\|)$. I.e., $R(u)$ is h-straight if, for all but finitely many v's, all (geodesic) continuations of $M(u,v)$ are contained in the cone $u + \mathcal{C}(v - u, h(\|v - u\|))$. For a given h, we define the (measurable) event

$$\Omega_1 = \{\omega \in \Omega_0 : \forall u \in \mathbb{Z}^{d*}, \ R(u) \text{ is } h\text{-straight }\}. \qquad (A.26)$$

In [N95], it was shown, under an exponential tail hypothesis on μ^* and another (plausible, but unfortunately unverified for any μ^*) hypothesis about asymptotic shape curvature, that for the h of (A.25), $\nu(\Omega_1) = 1$. With that as motivation, we give the following measurability result.

Proposition A.7 *For $d \geq 2$, the intersections of Ω_1 with $D_U(\hat{x})$, with $\{\exists$ disjoint $\hat{x}$-unigeodesics $\}$ and with $\{\exists$ an $(\hat{x}, \hat{y})$-bigeodesic $\}$ are all in $\mathcal{F}$.*

Proof. For $\omega \in \Omega_0$ and then for any $w_n \in \mathbb{Z}^{d*}$ with $||w_n|| \to \infty$, there exists a subsequence n_k such that

$$r_k = M(u, w_{n_k}) = (u, u_1^{(k)}, u_2^{(k)}, \ldots, u_{m_k}^{(k)} = w_{n_k}) \qquad (A.27)$$

converges to some unigeodesic

$$r_\infty = (u, u_1^{(\infty)}, u_2^{(\infty)}, \ldots). \qquad (A.28)$$

For $\omega \in \Omega_1$, it is easy to see (as noted in [N95]) that every unigeodesic is a $\hat{y}$-unigeodesic for some $\hat{y}$, and it is similarly easy to see that if the sequence w_n above is such that $w_n/||w_n|| \to \hat{x}$, then any subsequence limit r_∞ must be an $\hat{x}$-unigeodesic. Thus

$$\Omega_1 \cap A_{(u,v)}(\hat{x}) \;=\; \{\omega \in \Omega_1 : \exists \text{ an } \hat{x}\text{-unigeodesic starting with } (u, v)\} \qquad (A.29)$$

$$=\; \bigcap_{m=1}^{\infty} \{\omega \in \Omega_1 : \exists w \in \mathbb{Z}^{d*} \text{ with } ||w|| \geq m \text{ and } \theta(w, \hat{x}) \leq 1/m$$

$$\text{such that } M(u, w) \text{ starts with } (u, v)\},$$

which is in $\mathcal{F}$. The measurability of $\Omega_1 \cap D_U(\hat{x})$ follows, by the analogue of (A.9) with Ω_1 replacing Ω_0. Similarly $\Omega_1 \cap A_{u,u'}^*(\hat{x})$ and $\Omega_1 \cap \tilde{A}_u(\hat{x}, \hat{y})$ are in $\mathcal{F}$ (see (A.21)-(A.22) for definitions), because

$$\Omega_1 \cap A_{u,u'}^*(\hat{x}) \;=\; \bigcap_{m=1}^{\infty} \{\omega \in \Omega_1 : \; \exists w, w' \in \mathbb{Z}^{d*} \text{ with} \qquad (A.30)$$

$$||w||, ||w'|| \geq m \text{ and } \theta(w, \hat{x}), \theta(w', \hat{x}) \leq 1/m$$

$$\text{such that } M(u, w) \text{ and } M(u, w') \text{ are disjoint }\}$$

and

$$\Omega_1 \cap \tilde{A}_u(\hat{x}, \hat{y}) \;=\; \bigcap_{m=1}^{\infty} \{\omega \in \Omega_1 : \; \exists w, w' \in \mathbb{Z}^{d*} \text{ with} \qquad (A.31)$$

$$||w||, ||w'|| \geq m \text{ and } \theta(w, \hat{x}), \theta(w', \hat{y}) \leq 1/m$$

$$\text{such that } M(w, w') \text{ passes through } u\}.$$

The two remaining claims of the proposition easily follow, as in Proposition A.6.

$\square$

Remark A.8 For $\omega \in \Omega_0$, once all the finite geodesics, $M(u,v)$, have been defined, then the underlying $\tau(e^*)$'s play no role in any of the constructions or arguments of this appendix. Thus, our measurability results may be viewed from a more general perspective than that of first-passage percolation, as follows. Let $\tilde{\omega}$ denote any function from (unordered) pairs $\{u,v\}$ to (finite) paths $\tilde{\omega}_{\{u,v\}}$ $(= M(u,v))$ in $(\mathbb{Z}^{d*}, \mathbb{E}^{d*})$ connecting u and v. Let $\tilde{\Omega}$ be the space of all such $\tilde{\omega}$'s and let $\tilde{\mathcal{F}}$ be the σ-field generated by all subsets of the form $\{\tilde{\omega} \in \tilde{\Omega} : \tilde{\omega}_{\{u,v\}} = \tilde{r}\}$, for arbitrary pairs $\{u,v\}$ and finite paths $\tilde{r}$. Define $\tilde{\Omega}_0$ as

$$\tilde{\Omega}_0 = \Big\{ \tilde{\omega} \in \tilde{\Omega} : \forall \{u,v\}, \ \forall \{u',v'\}, \tag{A.32}$$

$$u' \text{ and } v' \text{ in } \tilde{\omega}_{\{u,v\}} \implies \tilde{\omega}_{\{u',v'\}} \text{ is a subpath of } \tilde{\omega}_{\{u,v\}} \Big\}.$$

Using the event $\tilde{\Omega}_0 \in \tilde{\mathcal{F}}$ as a substitute for the $\Omega_0 \in \mathcal{F}$ considered throughout this appendix, we may define, in an obvious way: $R(u)$ and $\hat{R}(u)$ for every u, the subsets of $\tilde{\Omega}_0$ analogous to all the subsets of Ω_0 considered in Propositions A.1 and A.6, the subset $\tilde{\Omega}_1$ of $\tilde{\Omega}_0$ analogous to Ω_1 defined in (A.26), and finally the subsets of $\tilde{\Omega}_1$ analogous to all the subsets of Ω_1 considered in Propostion A.7. The arguments presented in this appendix show that all these subsets are measurable, i.e., belong to $\tilde{\mathcal{F}}$.

Remark A.9 There are natural Euclidean models of first-passage percolation [VW92], [HoN96], based on homogeneous Poisson point processes in $\mathbb{R}^d$, where the deterministic graph $(\mathbb{Z}^{d*}, \mathbb{E}^{d*})$ is replaced by the "Poisson graph" (V^*, E^*), the random graph whose vertices are the Poisson "particle" locations and whose edges are all pairs $\{u,v\}$ of those locations. On the underlying probability space $(\Omega, \mathcal{F}, \nu)$ of these models, the finite geodesics $M(u,v)$ are (measurable) assignments, to each pair $\{u,v\}$ of particle locations, of a path on this graph connecting u and v. For some of these models, it can be shown that there is an $\Omega_0 \in \mathcal{F}$ with $\nu(\Omega_0) = 1$ such that on Ω_0 one has all the following properties:

1. V^* is infinite but locally finite, i.e., $V^* \cap \Lambda_L$ is finite for every L.

2. $\forall \{u,v\}, \ \{u',v'\} \in E^*, \ u' \text{ and } v' \in M(u,v) \implies M(u',v')$ is a subpath of $M(u,v)$.

3. $\forall L, \ \forall u \in V^* \cap \Lambda_L, \ \{v : \ M(u,v) \text{ is the single-edge path } \{u,v\}\}$ is finite.

4. If $\{u_1,v_1\}$ and $\{u_2,v_2\}$ in E^* have no vertices in common and for $j = 1,2$, $M(u_j,v_j)$ is the single-edge path $\{u_j,v_j\}$, then the two straight line segments in $\mathbb{R}^d$, $\overline{u_1,v_1}$ and $\overline{u_2,v_2}$, are disjoint.

For such models, all the measurability arguments and results of this appendix are valid.

Appendix B

Disordered Systems and Metastates

An infinite volume Gibbs state for couplings $J = (J_e : e \in \mathbb{E}^d)$ (and inverse temperature β, which will henceforth be fixed) is a probability measure on $\mathcal{S} = \{-1, +1\}^{\mathbb{Z}^d}$ (with the usual σ-field generated by cylinder sets) that satisfies extra requirements (the DLR equations). Let us denote by $\mathcal{M}(\mathcal{S})$ the set of all probability measures on $\mathcal{S}$. In this appendix, we give some technical lemmas and propositions about $\mathcal{M}(\mathcal{S})$-valued random variables Q and their distributions κ (on $\mathcal{M}(\mathcal{S})$) and, for disordered systems, about the joint distributions $\kappa^\dagger$ of (J, Q). Our primary interest is in $\kappa^\dagger$'s which are supported on configurations $(\mathcal{J}, q)$ such that q is an infinite volume Gibbs state for $\mathcal{J}$. Then the conditional distribution of Q given J will be a metastate $\kappa_{\mathcal{J}}$ — i.e., a probability measure on $\mathcal{M}(\mathcal{S})$ supported on the infinite volume Gibbs states for $\mathcal{J}$. We will begin however with more general Q's, κ's and $\kappa^\dagger$'s. It should be noted that the material presented here (other than Lemma B.5 and Proposition B.6) is basically contained, implicitly or explicitly, in the appendix of [AW90] and that this appendix, as a whole, is essentially identical to the appendix of [NS96d].

It is convenient to regard a probability measure on $\mathcal{S}$ as a point q in the product space $\Omega^1 \equiv \mathbb{R}^{\mathcal{A}}$, where

$$\mathcal{A} = \bigcup_{\Lambda \subset \mathbb{Z}^d \ finite} \left\{ (\Lambda, s) : s \in \mathcal{S}_\Lambda = \{-1, +1\}^\Lambda \right\}, \tag{B.1}$$

with the coordinate $q_{(\Lambda, s)}$ denoting the probability of the cylinder set $\{s' \in \mathcal{S} : s'_i = s_i \ \forall i \in \Lambda\}$. It is a standard fact that a point $q \in \Omega^1$ corresponds in this way to a probability measure if and only if it satisfies the three properties of positivity, normalization and consistency; i.e., for all finite subsets Λ, Λ' of $\mathbb{Z}^d$ with $\Lambda \subset \Lambda'$:

$$\forall s \in \mathcal{S}_\Lambda, \quad q_{(\Lambda, s)} \geq 0, \tag{B.2}$$

$$\sum_{s \in \mathcal{S}_\Lambda} q_{(\Lambda, s)} = 1, \tag{B.3}$$

$$\forall s \in \mathcal{S}_\Lambda, \quad q_{(\Lambda, s)} = \sum_{s' \in \mathcal{S}_{\Lambda'} : \ s' = s \ on \ \Lambda} q_{(\Lambda', s')}. \tag{B.4}$$

73

We will identify such a q with the (unique) corresponding probability measure on $\mathcal{S}$, and the set of all q's in Ω^1 satisfying (B.2)-(B.4) for all finite $\Lambda' \supset \Lambda$ will be identified with $\mathcal{M}(\mathcal{S})$. Clearly $\mathcal{M}(\mathcal{S})$ (now regarded as a subset of Ω^1) is measurable, i.e., it belongs to $\mathcal{F}^1$, the product σ-field over $\mathcal{A}$ of the Borel σ-field $\mathcal{B}$ on each $\mathbb{R}$.

Our first lemma uses elementary compactness arguments to construct, via an infinite volume limit, a joint distribution $\kappa^\dagger$ (for some random (J, Q)) on the product space

$$(\Omega^\dagger, \mathcal{F}^\dagger) = (\Omega^0, \mathcal{F}^0) \times (\Omega^1, \mathcal{F}^1), \tag{B.5}$$

where $(\Omega^0, \mathcal{F}^0)$ is the product over $\mathbb{E}^d$ of $(\mathbb{R}, \mathcal{B})$, such that $\kappa^\dagger$ is supported on $\{(\mathcal{J}, q) : q \in \mathcal{M}(\mathcal{S})\}$. For a sequence of subsets B_n of $\mathbb{E}^d$ we write $B_n \to \mathbb{E}^d$ if for every finite $B \subset \mathbb{E}^d$, $B_n \supset B$ for all large n; for subsets C_n of $\mathbb{Z}^d$ we similarly define $C_n \to \mathbb{Z}^d$.

Lemma B.1 *Let $B_n \subset \mathbb{E}^d$ and $C_n \subset \mathbb{Z}^d$ be finite and such that $B_n \to \mathbb{E}^d$ and $C_n \to \mathbb{Z}^d$. Suppose that for each n there are families of real valued random variables (on some $(\Omega, \mathcal{F}, \nu)$), $J^n = (J^n_e : e \in B_n)$ and $Q^n = (Q^n_{(\Lambda, s)} : \Lambda \subseteq C_n, s \in \mathcal{S}_\Lambda)$ such that*

$$\forall \text{ finite } \Lambda' \supset \Lambda, \ \forall \text{ large } n, \ q = Q^n \text{ satisfies (B.2)-(B.4)} \tag{B.6}$$

and

$$\forall e, \text{ the family of distributions of } J^n_e \text{ (as } n \text{ varies) is tight.} \tag{B.7}$$

Then the finite dimensional distributions of (J^n, Q^n) converge along a subsequence to a probability measure $\kappa^\dagger$ on $(\Omega^\dagger, \mathcal{F}^\dagger)$ with $\kappa^\dagger(\Omega^0 \times \mathcal{M}(\mathcal{S})) = 1$; i.e., there exists a subsequence n_k of the n's such that for any m, $m' \in \mathbb{Z}^+$, any $e_1, \ldots, e_m \in \mathbb{E}^d$, any $(\Lambda^1, s_1), \ldots, (\Lambda^{m'}, s_{m'}) \in \mathcal{A}$ and any bounded continuous real valued function f on $\mathbb{R}^{m+m'}$

$$E\left(f\left(J^{n_k}_{e_1}, \ldots, J^{n_k}_{e_m}, Q^{n_k}_{(\Lambda^1, s_1)}, \ldots, Q^{n_k}_{(\Lambda^{m'}, s_{m'})}\right)\right) \tag{B.8}$$

$$\to \int_{\Omega^\dagger} f\left(\mathcal{J}_{e_1}, \ldots, \mathcal{J}_{e_m}, q_{(\Lambda^1, s_1)}, \ldots, q_{(\Lambda^{m'}, s_{m'})}\right) d\kappa^\dagger((\mathcal{J}, q)).$$

Here, E denotes expectation with respect to ν.

Proof. For each L, and all sufficiently large n, let $\kappa^\dagger_{L,n}$ denote the joint distribution of the finitely many variables $J^n_{\{x,y\}}$, with x and y in the cube Λ_L, and $Q^n_{(\Lambda, s)}$ with $\Lambda \subseteq \Lambda_L$. The assumption (B.6) implies that the one-dimensional marginal of each $Q^n_{(\Lambda, s)}$ is supported on $[0, 1]$ and so the family of marginals, as n varies for a fixed (Λ, s), is tight; on the other hand, for a fixed e, the tightness of the J^n_e marginals was assumed. Thus as n varies for fixed L, the $\kappa^\dagger_{L,n}$'s are tight. It follows that we may choose a subsequence of the n's for each L (with the one for $L'' > L'$ a sub-subsequence of the one for L') along which $\kappa^\dagger_{L,n}$ converges to some probability measure $\kappa_L^\dagger$. By diagonalization, there is a single subsequence n_k so that for every L, $\kappa_{L,n_k}^\dagger \to \kappa_L^\dagger$ as $k \to \infty$.

Now for $L'' > L'$, the marginal of $\kappa_{L'',n_k}^\dagger$ for the variables coming from $\Lambda_{L'}$ (i.e., $J_{\{x,y\}}^{n_k}$ and $Q_{(\Lambda,s)}^{n_k}$ with $x, y \in \Lambda_{L'}$ and $\Lambda \subseteq \Lambda_{L'}$) is just $\kappa_{L',n_k}^\dagger$ and the same is true for $\kappa_{L''}^\dagger$ and $\kappa_{L'}^\dagger$. Thus the set of $\kappa_L^\dagger$'s is a consistent family of finite dimensional distributions corresponding to some $\kappa^\dagger$ on $(\Omega^\dagger, \mathcal{F}^\dagger)$. This proves (B.8).

It remains only to show that $\kappa^\dagger(\Omega^0 \times \mathcal{M}(\mathcal{S})) = 1$. But for any fixed finite $\Lambda' \supset \Lambda$, the set of q in Ω^1 satisfying (B.2)-(B.4) is a finite cylinder set corresponding to a *closed* subset $F_{\Lambda',\Lambda}$ of the finitely many variables appearing in (B.2)-(B.4). By assumption (B.6), this subset has $\kappa_{L,n}^\dagger$-probability one for all large L and n, and so, by the already proved convergence in distribution, the corresponding cylinder set has $\kappa^\dagger$-probability one. Since this is true for every finite $\Lambda' \supset \Lambda$, it follows that $\Omega^0 \times \mathcal{M}(\mathcal{S})$ has $\kappa^\dagger$-probability one, as desired. $\qquad\square$

In Chapter 4, we use the following special corollary of Lemma B.1, where $J_e^n = J_e$, with no dependence on n, and where Q^n is a functional of these fixed couplings. There Q^n is given by a Gibbs distribution on C_n for the fixed couplings.

Lemma B.2 *Let B_n and C_n be as in Lemma B.1 and let $(J_e : e \in \mathbb{E}^d)$ be a fixed family of random variables (on some $(\Omega, \mathcal{F}, \nu)$). Suppose for each n and each $\mathcal{J} \in \mathbb{R}^{B_n}$, $P_{\mathcal{J}}^n$ is a probability measure on $\mathcal{S}_{C_n}$ which depends measurably on $\mathcal{J}$, i.e., for each $s' \in \mathcal{S}_{C_n}$, $P_{\mathcal{J}}^n(\{s'\})$, as a function of $\mathcal{J} \in \mathbb{R}^{B_n}$, is Borel measurable. For $\Lambda \subseteq C_n$ and $s \in \mathcal{S}_\Lambda$, define random variables,*

$$Q_{(\Lambda,s)}^n = P_{(J_e : e \in B_n)}^n(\{t \in \mathcal{S}_{C_n} : t = s \text{ on } \Lambda\}); \tag{B.9}$$

then along some subsequence n_k of the n's,

$$(J, Q^{n_k}) = (J_e, Q_{(\Lambda,s)}^{n_k} : e \in \mathbb{E}^d, \Lambda \subseteq C_{n_k} \; s \in \mathcal{S}_\Lambda) \to \kappa^\dagger \tag{B.10}$$

in the sense of (B.8), where $\kappa^\dagger$ is some probability measure on $\Omega^0 \times \mathcal{M}(\mathcal{S})$ (i.e., $\kappa^\dagger$ is a probability measure on $(\Omega^\dagger, \mathcal{F}^\dagger)$ with $\kappa^\dagger(\Omega^0 \times \mathcal{P}(\mathcal{S})) = 1$).

Proof. This is an immediate corollary of Lemma B.1 by defining, for each n and each $e \in B_n$, $J_e^n = J_e$. $\qquad\square$

When B_n contains all $e = \{x, y\} \in \mathbb{E}^d$ with $x, y \in \Lambda_n$ and $P_{\mathcal{J}}^n$ is some Gibbs distribution on $\mathcal{S}_{C_n}$ (for the couplings $\mathcal{J}_e$ and the given β), then the probabilities $q_{(\Lambda,s)}^n \equiv P_{\mathcal{J}}^n(t = s \text{ on } \Lambda)$ satisfy the following finite volume version of the DLR equations: For $\Lambda, \Lambda' \subseteq C_n$ with $(\Lambda \cup \partial\Lambda) \subseteq \Lambda'$,

$$\forall s' \in \mathcal{S}_{\Lambda'}, \quad q_{(\Lambda',s')}^n = q_{(\Lambda' \setminus \Lambda, s'|_{\Lambda' \setminus \Lambda})}^n \; P_{\Lambda,\beta}^{s'|_{\partial\Lambda}}(\{s'|_\Lambda\}), \tag{B.11}$$

where $s'|_A$ denotes the restriction of s' to $A \subseteq \Lambda'$ and $P_{\Lambda,\beta}^{\bar{s}}$ denotes the usual finite volume Gibbs distribution on Λ for the couplings $\mathcal{J}_e$, with b.c. $\bar{s}$ on $\partial\Lambda$.

Lemma B.3 *Suppose that the hypotheses of Lemma B.2 are valid and, in addition, the probabilities $q_{(\Lambda,s)}^n \equiv P_{\mathcal{J}}^n(t = s \text{ on } \Lambda)$, satisfy (B.11) for all finite $\Lambda' \supseteq \Lambda \cup \partial\Lambda$ and all large n. Then the $\kappa^\dagger$ of Lemma B.2 has*

$$\kappa^\dagger\left(\{(\mathcal{J},q) : q \text{ is an infinite volume Gibbs distribution for } \mathcal{J}\}\right) = 1. \quad (B.12)$$

Proof. For a fixed Λ', Λ and s', the condition (B.11) is of the form $h = 0$ where h is a continuous function of $q_{(\Lambda',s')}^n$, $q_{(\Lambda'',s'')}^n$ (with $\Lambda'' = \Lambda' \setminus \Lambda$ and $s'' = s'|_{\Lambda'\setminus\Lambda}$) and $J_{e_1}, \ldots, J_{e_m}$ for certain fixed edges $e_1, \ldots, e_m$. Thus the set of these finitely many variables satisfying (B.11) is a closed subset $F_{\Lambda',\Lambda,s'}$. It follows from convergence in distribution (as in the proof of Lemma B.1) that the corresponding cylinder set has $\kappa^\dagger$-probability one. Thus the set H of $(\mathcal{J},q)$ such that $q \in \mathcal{M}(\mathcal{S})$ and (B.11) (with q^n replaced by q) is valid for all finite $\Lambda' \supseteq \Lambda\cup\partial\Lambda$ has $\kappa^\dagger(H) = 1$.

We consider a pair $(\mathcal{J}, P) \in H$. For a finite Λ and for L large enough so that the cube Λ_L contains $\Lambda \cup \partial\Lambda$, we denote by $\mathcal{F}_{\Lambda_L\setminus\Lambda}$ the σ-field generated by $\{s_x : x \in \Lambda_L \setminus \Lambda\}$. The conditions (B.11) imply that the conditional probability of P with respect to $\mathcal{F}_{\Lambda_L\setminus\Lambda}$ satisfies, for any $t \in \mathcal{S}_\Lambda$,

$$P\left(\{s \in \mathcal{S} : s = t \text{ on } \Lambda\}|\mathcal{F}_{\Lambda_L\setminus\Lambda}\right) = P_{\Lambda,\beta}^{\bar{s}}(t), \quad (B.13)$$

where $\bar{s} \in \mathcal{S}_{\partial\Lambda}$ is given by the (conditioned) values of s_x for $x \in \partial\Lambda$. By the martingale convergence theorem, we may let $L \to \infty$ to get

$$P\left(\{s \in \mathcal{S} : s = t \text{ on } \Lambda\}|\mathcal{F}_{\mathbb{Z}^d\setminus\Lambda}\right) = P_{\Lambda,\beta}^{\bar{s}}(t). \quad (B.14)$$

But the set of these equations for arbitrary finite Λ and $t \in \mathcal{S}_\Lambda$ is just the set of DLR equations which characterize P as an infinite volume Gibbs distribution for the couplings $(\mathcal{J}_e : e \in \mathbb{E}^d)$. Since the DLR equations also imply the validity of all the conditions (B.11), we see that the set of $(\mathcal{J},q)$ appearing in (B.12) is identical to H. Since $H \in \mathcal{F}^\dagger$ with $\kappa^\dagger(H) = 1$, the proof is complete. $\square$

For any probability measure $\kappa^\dagger$ on $(\Omega^\dagger, \mathcal{F}^\dagger) = (\Omega^0, \mathcal{F}^0) \times (\Omega^1, \mathcal{F}^1)$, one can define the marginal distribution (of $\mathcal{J}$), $\bar{\nu}$, on $(\Omega^0, \mathcal{F}^0)$ and the conditional distribution (of q, given $\mathcal{J}$), $\kappa_{\mathcal{J}}$, which for $\bar{\nu}$-a.e. $\mathcal{J}$ is a probability measure on $(\Omega^1, \mathcal{F}^1)$. For any $F \in \mathcal{F}^\dagger$,

$$\kappa^\dagger(F) = \bar{E}[\kappa_{\mathcal{J}}(F_{\mathcal{J}})], \quad (B.15)$$

where $\bar{E}$ denotes expectation with respect to $\bar{\nu}$ and

$$F_{\mathcal{J}} = \{q \in \Omega^1 : (\mathcal{J},q) \in F\}. \quad (B.16)$$

As a corollary of Lemma B.3, we have the following.

Lemma B.4 *Let $\kappa_{\mathcal{J}}$ be the conditional distribution of q, given $\mathcal{J}$, for the $\kappa^\dagger$ of Lemma B.2. Under the hypotheses of Lemma B.3,*

$$\text{for } \bar{\nu}\text{-a.e. } \mathcal{J}, \quad \kappa_{\mathcal{J}}\left(\{q : q \text{ is an infinite volume Gibbs distribution for } \mathcal{J}\}\right) = 1. \quad (B.17)$$

Proof. This is an immediate consequence of (B.15) by taking F to be the event appearing in (B.12). $\square$

The rest of the appendix is concerned, in the context of Lemma B.2, with the convergence of empirical distributions to $\kappa_{\mathcal{J}}$. For each $(\Lambda, s) \in \mathcal{A}$, $Q^n_{(\Lambda,s)}$ is a random variable, defined for $\{n : \Lambda \subseteq C_n\}$ and thus for all large n; we make the convention that $Q^n_{(\Lambda,s)} = 0$ for those finitely many n's for which $\Lambda \not\subseteq C_n$. We consider, for each L, the finite dimensional random vector

$$\vec{Q}^n_L = \left(Q^n_{(\Lambda,s)} : \Lambda \subseteq \Lambda_L, \ s \in \mathcal{S}_\Lambda \right), \tag{B.18}$$

and the empirical distributions, along some subsequence n'_k,

$$\kappa^{L,K} = \frac{1}{K} \sum_{k=1}^{K} \delta_{\vec{Q}^{n'_k}_L}. \tag{B.19}$$

These are (random) probability measures on $\mathbb{R}^{m_L}$, for some finite m_L.

In the context of Lemma B.2, $\kappa^{L,K}$ is random because it depends (measurably) on J. To denote this dependence, we will sometimes write $\kappa^{L,K}_{\mathcal{J}}$. We are interested in whether and in what sense $\kappa^{L,K}$ converges as $K \to \infty$ to some limit (random) probability measure on $\mathbb{R}^{m_L}$. We note that there is a natural candidate for the limit measure - namely, take the conditional distribution $\kappa_{\mathcal{J}}$, of q, given $\mathcal{J}$, for the joint distribution $\kappa^\dagger$ and consider its marginal distribution $\kappa^L_{\mathcal{J}}$ for $\left(q_{(\Lambda,s)} : \Lambda \subseteq \Lambda_L, \ s \in \mathcal{S}_\Lambda \right)$. The next lemma is a technical result which leads to the subsequent proposition giving a.s. convergence to this limit along a subsequence.

Lemma B.5 *Assume the hypotheses of Lemma B.2. For any fixed L and fixed $\vec{r} \in \mathbb{R}^{m_L}$, there exists a sub-subsequence n'_k of the subsequence n_k of Lemma B.2, such that (with $\kappa^{L,K}$ given by (B.18)-(B.19)), the sequence of random variables,*

$$\Phi^K_{L,\vec{r}} = \int_{\mathbb{R}^{m_L}} e^{i\,(\vec{r},\vec{q}_L)} d\kappa^{L,K}(\vec{q}_L) = \frac{1}{K} \sum_{k=1}^{K} e^{i\,(\vec{r},\vec{Q}^{n'_k}_L)}, \tag{B.20}$$

(where $(\vec{r}, \vec{q}_L)$ denotes the standard inner product on $\mathbb{R}^{m_L}$) converges in $L^2(\Omega, \mathcal{F}, \nu)$, as $K \to \infty$, to some random variable $\Phi_{L,\vec{r}}$.

Proof. Let $Y_n = e^{i\,(\vec{r},\vec{Q}^n_L)}$. We must show that n'_k can be chosen so that

$$\begin{aligned}
E\left(\Phi^{K'}_{L,\vec{r}} - \Phi^K_{L,\vec{r}} \right)^2 &= \frac{1}{K'^2} \sum_{k,j=1}^{K'} E\left(Y_{n'_k} Y_{n'_j} \right) + \frac{1}{K^2} \sum_{k,j=1}^{K} E\left(Y_{n'_k} Y_{n'_j} \right) \\
&\quad - \frac{2}{K'K} \sum_{k=1}^{K'} \sum_{j=1}^{K} E\left(Y_{n'_k} Y_{n'_j} \right) \to 0
\end{aligned} \tag{B.21}$$

as $K, K' \to \infty$. Note that the total contribution of the diagonal terms (where $k = j$) in the last expression is negligible as $K, K' \to \infty$, as is the contribution from terms with k (or j) fixed. It follows that it suffices to choose n'_k so that

$$\lim_{\substack{k,j\to\infty \\ k\neq j}} E\left(Y_{n'_k}Y_{n'_j}\right) \text{ exists.} \tag{B.22}$$

We will do this by a compactness argument.

We will use the facts that $C_{mn} \equiv E(Y_m Y_n)$ is bounded in modulus by 1 and that $C_{mn} = C_{nm}$. We begin by choosing, for each l, a subsequence of the n's (with the one for $l'' > l'$ a subsequence of the one for l') along which C_{ln} converges to some α_l. By diagonalization, there is a fixed subsequence $\tilde{n}_k$ so that, for every l, $C_{l\tilde{n}_k} \to \alpha_l$ as $k \to \infty$. Since the α_l's are also bounded in modulus by one, we may next pick a sub-subsequence n_k^* of $\tilde{n}_k$ so that $\alpha_{n_k^*} \to \alpha$. We now have:

$$\forall k, \quad C_{n_k^*, n_{k'}^*} \to \alpha_{n_k^*} \quad \text{as } k' \to \infty, \quad \text{and} \quad \alpha_{n_k^*} \to \alpha \quad \text{as } k \to \infty. \tag{B.23}$$

Finally, we define n'_k inductively by taking $n'_1 = n_1^*$ $(= n_{k_1}^*$ with $k_1 = 1)$ and, for $j \geq 1$, taking $n'_{j+1} = n_{k_{j+1}}^*$ with k_{j+1} the smallest $k' > k_j$ such that

$$\left|C_{n'_j n_k^*} - \alpha_{n'_j}\right| < \frac{1}{2^j} \quad \text{for } k \geq k'. \tag{B.24}$$

It follows that

$$\limsup_{\substack{j\to\infty \\ k>j}} \left|C_{n'_j n'_k} - \alpha\right| \leq \lim_{j\to\infty} \left(\frac{1}{2^j} + \left|\alpha_{n'_j} - \alpha\right|\right) = 0. \tag{B.25}$$

But this, together with $C_{mn} = C_{nm}$, implies (B.22). $\square$

Proposition B.6 *Assume the hypotheses of Lemma B.2. Then there exists a sub-subsequence n'_k of the subsequence n_k of Lemma B.2 and some subsequence K_m of the K's, such that, almost surely, the empirical distributions κ^{L,K_m} (given by (B.18)-(B.19)) converge (weakly) as $m \to \infty$ to the finite dimensional marginals κ_J^L (of the conditional distribution $\kappa_{\mathcal{J}}$ of the $\kappa^\dagger$ of Lemma B.2, with $\mathcal{J} = J$). I.e., for ν-a.e. ω, $\forall L$, $\forall$ bounded continuous functions f on $\mathbb{R}^{m_L}$,*

$$\lim_{m\to\infty} \int_{\mathbb{R}^{m_L}} f(\vec{q}_L) d\kappa^{L,K_m}_{J(\omega)}(\vec{q}_L) = \int_{\Omega^1} f\left((q_{(\Lambda,s)} : \Lambda \subseteq \Lambda_L, \ s \in \mathcal{S}_\Lambda)\right) d\kappa_{J(\omega)}(q). \tag{B.26}$$

Proof. By Lemma B.5 and a diagonalization argument, we can replace the n'_k of Lemma B.5 by a single further sub-subsequence (which we will still call n'_k) so that $\Phi^K_{L,\vec{r}}$ converges in L^2 to some $\Phi_{L,\vec{r}}$, as $K \to \infty$, for every L and every $\vec{r} \in \mathbb{R}^{m_L}$ with rational coordinates. But L^2 convergence implies convergence in probability, which implies a.s. convergence along a subsequence of K's. Thus, by a further diagonalization, we have for a single subsequence K_m of the K's that $\Phi^{K_m}_{L,\vec{r}}$ converges a.s. to $\Phi_{L,\vec{r}}$, as $m \to \infty$, for all L and rational $\vec{r}$.

Now for each ω, $\Phi_{L,\vec{r}}^K$, as a function of $\vec{r}$, is the characteristic function of a probability measure $\kappa_{J(\omega)}^{L,K}$ supported on the compact set $[0,1]^{m_L}$ (since each $Q_{(\Lambda,s)}^n$ takes values in $[0,1]$). The tightness of this family of measures (as K varies) along with convergence (for $K = K_m$) of the characteristic functions for a dense set of $\vec{r}$'s implies that for ν-a.e. ω,

$$\forall L, \ \forall \vec{r} \in \mathbb{R}^{m_L}, \quad \Phi_{L,\vec{r}}^{K_m} \to \Phi_{L,\vec{r}} = \int_{\mathbb{R}^{m_L}} e^{i\,(\vec{r},\vec{q}_L)} d\kappa^L(\vec{q}_L) \quad \text{as } m \to \infty, \quad (B.27)$$

where κ^L is some probability measure on $\mathbb{R}^{m_L}$, depending on ω.

To complete the proof of the proposition, it only remains to identify κ^L with the marginal distribution of $\kappa_{J(\omega)}$. Because of the relation between $\kappa_{\mathcal{J}}$ and $\kappa^\dagger$, to do this, it suffices to show that for every bounded continuous function $g(\mathcal{J})$, depending on only finitely many $\mathcal{J}_e$'s,

$$E\left(g(J)\Phi_{L,\vec{r}}\right) = \int g(\mathcal{J}) e^{i\,(\vec{r},\vec{q}_L)} d\kappa^\dagger((\mathcal{J},q)). \quad (B.28)$$

But by (B.20) and Lemma B.2

$$E\left(g(J)\Phi_{L,\vec{r}}\right) = \lim_{m \to \infty} E\left(g(J)\Phi_{L,\vec{r}}^{K_m}\right) = \lim_{m \to \infty} \frac{1}{K_m} \sum_{k=1}^{K_m} E\left(g(J)e^{i\,(\vec{r},\vec{Q}_L^{n'_k})}\right)$$

$$= \int_{\Omega^\dagger} g(\mathcal{J}) e^{i\,(\vec{r},\vec{q}_L)} d\kappa^\dagger((\mathcal{J},q)), \quad (B.29)$$

as desired. $\qquad\qquad\qquad\qquad\qquad\qquad\qquad\qquad\qquad\qquad\qquad\qquad\qquad\qquad\square$

Bibliography

[A80] M. Aizenman, Translation invariance and instability of phase coexistence in the two dimensional Ising System, Commun. Math. Phys. 73, 1980, pp. 83–94.

[A97] M. Aizenman, On the number of incipient spanning clusters, Nucl. Phys. B[FS] 485, 1997, pp. 551–582.

[ACCN87] M. Aizenman, J. Chayes, L. Chayes, C.M. Newman, The phase boundary in dilute and random Ising and Potts ferromagnets, J. of Physics A: Mathematical and General 20, 1987, pp. 313–318.

[ACCN88] M. Aizenman, J. Chayes, L. Chayes, C.M. Newman, Discontinuity of the magnetization in one-dimensional $1/|x - y|^2$ Ising and Potts models, J. of Stat. Phys. 50, 1988, pp. 1–40.

[AKN87] M. Aizenman, H. Kesten and C.M. Newman, Uniqueness of the infinite cluster and continuity of connectivity functions for short and long range percolation, Commun. Math. Phys. 111, 1987, pp. 505–531.

[AW90] M. Aizenman and J. Wehr, Rounding effects of quenched randomness of first-order phase transitions, Commun. Math. Phys. 130, 1990, pp. 489–528.

[AlS92] D. Aldous and J.M. Steele, Asymptotics for Euclidean minimal spanning trees on random points, Prob. Th. Rel. Fields 92, 1992, pp. 247–258.

[Ale95] K. Alexander, Percolation and minimal spanning forests in infinite graphs, Ann. Prob. 23, 1995, pp. 87–104.

[AleC96] K. Alexander and L. Chayes, Non-perturbative criteria for Gibbsian uniqueness, 1996 preprint, Commun. Math. Phys. (to appear).

[AmPZ92] J.M.G. Amaro de Matos, A.E. Patrick and V.A. Zagrebnov, Random infinite-volume Gibbs states for the Curie-Weiss random field Ising model, J. Stat. Phys. 66, 1992, pp. 139–164.

[BD86] L.A. Bassalygo and R.L. Dobrushin, Uniqueness of a Gibbs field with random potential – an elementary approach, Theory Prob. Appl. 31, 1986, pp. 572–589.

[BM94] J. van den Berg and C. Maes, Disagreement percolation in the study of of Markov fields, Ann. Prob. 22, No. 2, 1994, pp. 749–763.

[Bo96] D. Boivin, Ergodic theorems for surfaces with minimal random weights, 1996 preprint.

[BoCKS97] C. Borgs, J. Chayes, H. Kesten and J. Spencer, The birth of the infinite cluster: finite-size scaling in percolation, in preparation.

[BoF86] A. Bovier and J. Fröhlich, A heuristic theory of the spin glass phase, J. Stat. Phys. 44, 1986, pp. 347–391.

[BoK94] A. Bovier and C. Külske, A rigorous renormalization group method for interfaces in random media, Rev. Math. Phys. 6, 1994, pp. 413–496.

[BoK96] A. Bovier and C. Külske, There are no nice interfaces in $(2 + 1)$-dimensional SOS models in random media, J. Stat. Phys. 83, 1996, pp.751–759.

[BoP97] A. Bovier and P. Picco (eds.), Mathematics of Spin Glasses and Neural Networks, Birkhäuser, Boston, 1997 (to appear).

[BrM87] A.J. Bray and M.A. Moore, Chaotic nature of the spin-glass phase, Phys. Rev. Lett. 58, 1987, pp. 57–60.

[BroH57] S.R. Broadbent and J.M. Hammersley, Percolation processes I. Crystals and mazes, Proc. Camb. Phil. Soc. 53, 1957, pp. 629–641.

[BuK89] R. Burton and M. Keane, Density and uniqueness in percolation, Commun. Math. Phys. 121, 1989, pp. 501–505.

[CKLW82] R. Chandler, J. Koplik, K. Lerman and J.F. Willemsen, Capillary displacement and percolation in porous media, J. Fluid Mech. 119, 1982, pp. 249–267.

[ChCN85] J. Chayes, L. Chayes and C.M. Newman, The stochastic geometry of invasion percolation, Commun. Math. Phys. 101, 1985, pp. 383–407.

[CiMB94] M. Cieplak, A. Maritan and J.R. Banavar, Optimal paths and domain walls in the strong disorder limit, Phys. Rev. Lett. 72, 1994, pp. 2320–2323.

[Co89] F. Comets, Large deviation estimates for a conditional probability distribution. Applications to random interaction Gibbs measures, Prob. Theory Rel. Fields 80, 1989, pp. 407–432.

[DG96] E. De Santis and A. Gandolfi, Bond percolation in frustrated systems, in preparation, and private communication, June, 1996.

[Do68] R.L. Dobrushin, The description of a random field by means of conditional probabilities and conditions of its regularity, Theory of Prob. and its Appl. 13, 1968, pp. 197–224.

[DoS85] R.L. Dobrushin and S.B. Shlosman, Constructive criteria for the uniqueness of a Gibbs field, pp. 371–403, in Statistical Mechanics and Dynamical Systems, J. Fritz, A. Jaffe and D. Szász (eds.), Birkhäuser, Boston, 1985.

[DrKP95] H. von Dreifus, A. Klein and J. Fernando Perez, Taming Griffiths' singularities: infinite differentiability of quenched correlation functions, Commun. Math. Phys. 170, 1995, pp. 21–39.

[EA75] S.F. Edwards and P.W. Anderson, Theory of spin glasses, J. Phys. F 5, 1975, pp. 965–974.

[ES88] R.G. Edwards and A.D. Sokal, Generalization of the Fortuin-Kasteleyn-Swendsen-Wang representation and Monte Carlo algorithm, The Physical Review D 38, 1988, pp. 2009–2012.

[En90] A.C.D. van Enter, Stiffness exponent, number of pure states, and Almeida-Thouless line in spin-glasses, J. Stat. Phys. 60, 1990, pp. 275–279.

[EnHM92] A.C.D. van Enter, A. Hof and J. Miękisz, Overlap distributions for deterministic systems with many pure states, J. Phys. A 25, 1992, pp. L1133–L1137.

[FH86] D.S. Fisher and D.A. Huse, Ordered phase of short-range Ising spin-glasses, Phys. Rev. Lett. 56, 1986, pp. 1601–1604.

[FH88] D.S. Fisher and D.A. Huse, Equilibrium behavior of the spin-glass ordered phase, Phys. Rev. B 38, 1988, pp. 386–411.

[FS90] M.E. Fisher and R.R.P. Singh, Critical Points, large-dimensionality expansions and Ising spin glass, pp. 87–111, in Disorder in physical systems, G.R. Grimmett and D.J.A. Welsh (eds.), Clarendon Press, Oxford, 1990.

[FoLN91] G. Forgacs, R. Lipowsky and T.M. Nieuwenhuizen, The behaviour of interfaces in ordered and disordered systems, pp. 135–363, in Phase Transitions and Critical Phenomena, C. Comb and J. Lebowitz (eds.), Vol. 14, Acad. Press, London, 1991.

[For72] C.M. Fortuin, On the random-cluster model. III. The simple random-cluster model, Physica 59, 1972, pp. 545–570.

[ForK72] C.M. Fortuin and P.W. Kasteleyn, On the random-cluster model. I. Introduction and relation to other models, Physica 57, 1972, pp. 536–564.

[ForKG71] C.M. Fortuin, P.W. Kasteleyn and J. Ginibre, Correlation inequalities on some partially ordered set, Commun. Math. Phys. 22, 1971, pp. 89–103.

[GKN92] A. Gandolfi, M. Keane and C.M. Newman, Uniqueness of the infinite component in a random graph with applications to percolation and spin glasses, Prob. Theory Rel. Fields 92, 1992, pp. 511–527.

[Ge88] H.O. Georgii, Gibbs Measures and Phase Transitions, de Gruyter Studies in Math., Bd. 9, Berlin, 1988.

[GiM95] G. Gielis and C. Maes, The uniqueness regime of Gibbs fields with unbounded disorder, J. Stat. Phys. 81, 1995, pp. 829–835.

[Gr67] R.B. Griffiths, Correlation in Ising ferromagnets I and II, J. Math. Phys. 8, 1967, pp. 478–489.

[Gri94] G.R. Grimmett, Percolative problems, pp. 69–86, in Probability and Phase Transition, G. Grimmett (ed.), Kluwer, Dordrecht, 1994.

[Gu95] F. Guerra, private communication, September, 1995.

[H96] O. Häggström, Random-cluster representations in the study of phase transitions, 1996 preprint.

[Ha57] J.M. Hammersley, Percolation processes. Lower bounds for the critical probabiliy, Ann. Math. Stat. 28, 1957, pp. 790–795.

[Ha59] J.M. Hammersley, Bornes supérieures de la probabilité critique dans un processus de filtration, pp. 17–37, in Le Calcul des Probabilités et ses Applications, CNRS, Paris, 1959.

[HaW65] J.M. Hammersley and D.J.A. Welsh, First-passage percolation, subadditive processes, stochastic networks and generalized renewal theory, pp. 61–110, in Bernoulli, Bayes, Laplace Anniversary Volume, J. Neyman and L. Lecam (eds.), Springer-Verlag, Berlin, 1965.

[HarS90] T. Hara and G. Slade, Mean-field critical behavior for percolation in high dimensions, Commun. Math. Phys. 128, 1990, pp. 333–391.

[Harr60] T.E. Harris, A lower bound for the critical probability in a certain percolation process, Proc. Camb. Philosophical Soc. 56, 1960, pp. 13–20.

[Hi81] Y. Higuchi, On the absence of non-translation invariant Gibbs states for the two-dimensional Ising model, Vol. I, pp. 517–534, in Random Fields, Esztergom (Hungary) 1979, J. Fritz, J.L. Lebowitz and D. Szász (eds.), Amsterdam, North Holland.

[HoJY83] A. Houghton, S. Jain and A.P. Young, Role of initial conditions in spin glass dynamics and significance of Parisi's $q(x)$, J. Phys. C 16, 1983, pp. L375–L381.

[HoN96] C.D. Howard and C.M. Newman, Euclidean models of first-passage percolation, 1996 preprint, Prob. Theory Rel. Fields (to appear).

[HuF87] D.A. Huse and D.S. Fisher, Pure states in spin glasses, J. Phys. A 20, 1987, pp. L997–L1003.

[HuH85] D.A. Huse and C.L. Henley, Pinning and roughening of domain walls in Ising systems due to random impurities, Phys. Rev. Lett. 54, 1985, pp. 2708–2711.

[HuHF85] D.A. Huse, C.L. Henley and D.S. Fisher, Respond, Phys. Rev. Lett. 55, 1985, p. 2924.

[K85] M. Kardar, Roughening by impurities at finite temperature, Phys. Rev. Lett. 55, 1985, p. 2923.

[KPZ86] M. Kardar, G. Parisi and Y.-C. Zhang, Dynamic scaling of growing interfaces, Phys. Rev. Lett. 56, 1986, pp. 889–892.

[KaO88] Y. Kasai and A. Okiji, Percolation problem describing $\pm J$ Ising spin glass system, Progress in Theoretical Physics 79, 1988, pp. 1080–1094.

[KasF69] P.W. Kasteleyn and C.M. Fortuin, Phase transitions in lattice systems with random local properties, Journal of the Physical Society of Japan, 26, 1969, pp. 11–14.

[KeS68] D.G. Kelly and S. Sherman, General Griffiths' inequalities on correlations in Ising ferromagnets, J. Math. Phys. 9, 1968, pp. 466–484.

[Kes80] H. Kesten, The critical probability of bond percolation on the square lattice equals 1/2, Commun. Math. Phys. 74, 1980, pp. 41–59.

[Kes87] H. Kesten, Surfaces with minimal random weights and maximal flows. A higher dimensional version of first-passage percolation, Illinois J. Math. 31, pp. 91–166.

[Ku96] C. Külske, Metastates in disordered mean field models: random field and Hopfield models, 1996 preprint, to appear in [BoP97].

[LM72] J.L. Lebowitz and A. Martin-Löf, On the uniqueness of the equilibrium state for Ising spin systems, Commun. Math. Phys. 25, 1972, pp. 276–282.

[Le77] F. Ledrappier, Pressure and variational principle for random Ising model, Commun. Math. Phys. 56, 1977, pp. 297–302.

[LeB80] R. Lenormand and S. Bories, Description d'un mécanisme de connexion de liaison destiné à l'étude du drainage avec piègeage en milieu poreux, C. R. Acad. Sci. Paris Sér. B 291, 1980, pp. 279–282.

[LiN96] C. Licea and C.M. Newman, Geodesics in two-dimensional first-passage percolation, Ann. Prob. 24, 1996, pp. 399–410.

[LiNP96] C. Licea, C.M. Newman and M.S.T Piza, Superdiffusivity in first-passage percolation, Prob. Theory Rel. Fields 106, 1996, pp. 559–591.

[M84] W.L. McMillan, Scaling theory of Ising spin glasses, J. Phys. C 17, 1984, pp. 3179–3187.

[MPSTV84] M. Mézard, G. Parisi, N. Sourlas, G. Toulouse and M.A. Virasoro, Nature of spin-glass phase, Phys. Rev. Lett. 52, 1984, pp. 1156–1159.

[MPV87] M. Mézard, G. Parisi and M.A. Virasoro, Spin Glass Theory and Beyond, World Scientific, Singapore, 1987.

[N91] C.M. Newman, Ising models and dependent percolation, pp. 395–401, in Topics in Statistical Dependence, H.W. Block, A.R. Sampson and T.H Savits (eds.), IMS Lecture Notes – Monograph Series 16, 1991.

[N94] C.M. Newman, Disordered Ising systems and random cluster representation, pp. 247–260, in Probability and Phase Transition, G. Grimmett (ed.), Kluwer, Dordrecht, 1994.

[N95] C.M. Newman, A surface view of first-passage percolation, pp. 1017–1023, in Proceedings of the International Congress of Mathematicians, S.D. Chatterji (ed.), Birkhäuser Verlag, Basel, 1995.

[NS92] C.M. Newman and D.L. Stein, Multiple states and thermodynamic limits in short ranged Ising spin glass models, Phys. Rev. B 46, 1992, pp. 973–982.

[NS94] C.M. Newman and D.L. Stein, Spin glass model with dimension-dependent ground state multiplicity, Phys. Rev. Lett. 72, 1994, pp. 2286–2289.

[NS96a] C.M. Newman and D.L. Stein, Ground-state structure in a highly disordered spin-glass model, J. Stat. Phys. 82, 1996, pp. 1113–1132.

[NS96b] C.M. Newman and D.L. Stein, Non-mean-field behavior of realistic spin glass, Phys. Rev. Lett. 76, 1996, pp. 515–518.

[NS96c] C.M. Newman and D.L. Stein, Spatial inhomogeneity and thermodynamic chaos, Phys. Rev. Lett. 76, 1996, pp. 4821–4824.

[NS96d] C.M. Newman and D.L. Stein, Thermodynamic chaos and the structure of short-range spin glasses, 1996 preprint, to appear in [BoP97].

[NS97] C.M. Newman and D.L. Stein, The metastate approach to thermodynamic chaos, Phys. Rev. E 55, 1997, pp. 5194–5211.

[P79] G. Parisi, Infinite number of order parameters for spin-glasses, Phys. Rev. Lett. 43, 1979, pp. 1754–1756.

[P83] G. Parisi, Order parameter for spin-glasses, Phys. Rev. Lett. 50, 1983, pp. 1946–1948.

[PS91] L.A. Pastur and M.V. Shcherbina, Absence of self-averaging of the order parameter in the Sherrington-Kirkpatrick model, J. Stat. Phys. 62, 1991, pp. 1–19.

[PST97] L. Pastur, M. Shcherbina and B. Tirozzi, Disordered systems: self-averaging and replica symmetric solution, 1997 preprint.

[Pi96] A. Pisztora, Surface order large deviations for Ising, Potts and percolation models, Prob. Theory Rel. Fields 104, 1996, pp. 427–466.

[S69]   T.K. Sarkar, Some lower bound of reliability, Technical Report No. 124, Dept. of Operations Research and Statistics, Standford University, Stanford, CA, 1969.

[S95]   T. Seppäläinen, Entropy, limit theorems and variational principle for disordered lattice systems, Commun. Math. Phys. 171, 1995, pp. 233–277.

[SK75] D. Sherrington and S. Kirkpatrick, Solvable model of a spin glass, Phys. Rev. Lett. 35, 1975, pp. 1792–1796.

[SW87] R.H. Swendsen and J.S. Wang, Nonuniversal critical dynamics in Monte Carlo simulations, Phys. Rev. Lett. 58, 1987, pp. 86–88.

[T96]   M. Talagrand, The Sherrington-Kirkpatrick model: a challenge for mathematicians, 1996 preprint, Prob. Theory Rel. Fields (to appear).

[VW92] M.Q. Vahidi-Asl and J.C. Wierman, A shape result for first-passage percolation on the Voronoi tesselation and Delaunay triangulation, pp. 247–262, in Random Graphs '89, A. Frieze and T. Łuczak (eds.), Wiley, New York, 1992.

[W96]   J. Wehr, On the number of infinite geodesics and ground states in disordered systems, 1996 preprint.

[WW96] J. Wehr and J. Woo, Absence of geodesics in first passage percolation on a half-plane, 1996 preprint.

[WiW83] D. Wilkinson and J.F. Willemsen, Invasion percolation: A new form of percolation theory, J. Phys. A 16, 1983, pp. 3365–3376.

[Z96]   M.P.W. Zerner, private communication, April, 1996.

Index